THRESHOLD

**Cambridge Pre-GED Program
in
Mathematics

2**

THRESHOLD

Cambridge Pre-GED Program in Mathematics

2

CAMBRIDGE Adult Education
REGENTS/PRENTICE HALL
Englewood Cliffs, New Jersey 07632

Library of Congress Cataloging-in-Publication Data

Threshold : Cambridge pre-GED program in mathematics 2.
    p.   cm.
  "Cambridge Adult Education."
  ISBN 0-13-917600-4 (pbk.)
  1. Fractions.   2. Decimal fractions.   3. Graphic methods.
  4. Elementary education of adults.   I. Cambridge Adult Education
(Firm)
QA117.T47 1992
513.2'6--dc20   92-19418
                                                              CIP

Publisher: **TINA B. CARVER**
Executive Editor: **JAMES W. BROWN**
Editorial Supervisor: **TIMOTHY A. FOOTE**
Managing Editor: **SYLVIA MOORE**
Production Editor: **JANET S. JOHNSTON**
Copyeditor: **RICHARD CONNER**
Pre-press Buyer: **RAY KEATING**
Manufacturing Buyer: **LORI BULWIN**
Scheduler: **LESLIE COWARD**
Interior designers: **JANET SCHMID** and **JANET S. JOHNSTON**
Cover coordinator: **MARIANNE FRASCO**
Cover designer: **BRUCE KENSELAAR**
Cover photo: The Oregon Coast. **CRAIG TUTTLE/THE STOCK MARKET®**
Photo researcher: **ELLEN GRATKOWSKI**

© 1993 by REGENTS/PRENTICE HALL
A Division of Simon & Schuster
Englewood Cliffs, New Jersey 07632

All rights reserved. No part of this book may be
reproduced, in any form or by any means, without
permission in writing from the publisher.

Printed in the United States of America

10 9 8 7 6 5 4 3

ISBN 0-13-917600-4

PHOTO CREDITS
Page 1 Hakim Raquib
        9 Irene Springer
   109 Ken Karp
   171 IBM Corporation
   211 Hakim Raquib

Prentice-Hall International (UK) Limited, *London*
Prentice-Hall of Australia Pty. Limited, *Sydney*
Prentice-Hall Canada Inc., *Toronto*
Prentice-Hall Hispanoamericana, S.A., *Mexico*
Prentice-Hall of India Private Limited, *New Delhi*
Prentice-Hall of Japan, Inc., *Tokyo*
Simon & Schuster Asia Pte. Ltd., *Singapore*
Editora Prentice-Hall do Brasil, Ltda., *Rio de Janeiro*

CONTENTS

ACKNOWLEDGMENTS   xi

INTRODUCTION   xiii

TO THE STUDENT   xv

PRETEST   *1*

    Fraction Problems, 2
    Percent Problems, 4
    Problems Based on Tables and Graphs, 5

UNIT 1   FRACTIONS   *9*

    Chapter 1   UNDERSTANDING FRACTIONS   10

        Lesson 1   Proper Fractions, 10
          Writing Proper Fractions, 11

        Lesson 2   Simplifying Proper Fractions, 13
          Writing Fractions in Lower Terms, 14
          Simplifying Fractions to Lowest Terms, 15

        Lesson 3   Raising Proper Fractions to Higher Terms, 16
          Multiplying by Any Number to Raise Terms, 17
          Multiplying by a Certain Number to Raise Terms, 18

        Lesson 4   Finding Lowest Common Denominators, 20
          Using Multiples to Find the LCD, 20
          Dividing Denominators to Find the LCD, 21
          Multiplying Denominators to Find the LCD, 22

        Lesson 5   Comparing and Ordering Fractions, 23
          Comparing Fractions with Unlike Denominators, 24
          Ordering Fractions with Unlike Denominators, 25

        Lesson 6   Improper Fractions and Mixed Numbers, 26
          Writing Improper Fractions, 26
          Writing Mixed Numbers, 29
          Rewriting Improper Fractions as Mixed or Whole Numbers, 31
          Rewriting Mixed Numbers as Improper Fractions, 32

Chapter 2 ADDITION   34

    Lesson 7   Adding Fractions with Like Denominators, 34

    Lesson 8   Adding Mixed Numbers with Like Denominators, 36
        Simplifying Sums with Fractions Smaller than One, 36
        Simplifying Sums with Fractions Equal to One, 37
        Simplifying Sums with Fractions Greater than One, 39

    Lesson 9   Adding Fractions with Unlike Denominators, 40

    Lesson 10   Adding Mixed Numbers with Unlike Denominators, 42

Chapter 3 SUBTRACTION   44

    Lesson 11   Subtracting with Like Denominators, 44
        Subtracting Fractions with Like Denominators, 44
        Subtracting Mixed Numbers with Like Denominators, 46

    Lesson 12   Subtracting with Unlike Denominators, 47
        Subtracting Fractions with Unlike Denominators, 47
        Subtracting Mixed Numbers with Unlike Denominators, 48

    Lesson 13   Borrowing, 51
        Subtracting Fractions from the Whole Number 1, 51
        Subtracting Fractions from Whole Numbers Larger than 1, 52
        Borrowing When Subtracting Mixed Numbers with Like Denominators, 53
        Borrowing When Subtracting Mixed Numbers with Unlike Denominators, 54

Mixed Practice 1   Addition and Subtraction of Fractions and Mixed Numbers   **56**

Chapter 4 MULTIPLICATION   58

    Lesson 14   Multiplying Fractions by Fractions, 58

    Lesson 15   Canceling before Multiplying Fractions, 59
        Canceling Once, 60
        Canceling More than Once, 61

    Lesson 16   Multiplying Whole Numbers by Fractions, 62

    Lesson 17   Multiplying Mixed Numbers by Fractions or Mixed Numbers, 64

Chapter 5 DIVISION   66

    Lesson 18   Dividing by Fractions, 66
        Dividing Fractions by Fractions, 68
        Dividing Whole Numbers or Mixed Numbers by Fractions, 68

    Lesson 19   Dividing Fractions or Mixed Numbers by Whole Numbers, 69
        Dividing Fractions by Whole Numbers, 69
        Dividing Mixed Numbers by Whole Numbers, 70

    Lesson 20   Dividing by Mixed Numbers, 72

Mixed Practice 2   Multiplication and Division of Fractions and Mixed Numbers 74

Fraction Skills Review   75

Chapter 6   USING FRACTIONS   77

    Lesson 21   Working with Measurements Expressed as Fractions or Mixed Numbers, 77
Converting Smaller Units of Measure to Larger Units, 78
Converting Larger Units of Measure to Smaller Units, 79
Converting Units When Two Conversions Are Involved, 80

    Lesson 22   Finding Perimeters, Circumferences, and Areas, 81
Finding Perimeters, 82
Finding Circumferences, 84
Finding Areas, 85

    Lesson 23   Working with Both Fractions and Decimals, 88
Rewriting Fractions as Decimals, 88
Rewriting Decimals as Fractions, 89
Solving Problems Containing a Fraction and a Decimal, 90

    Lesson 24   Rounding Mixed Numbers and Estimating Solutions, 91
Rounding Mixed Numbers, 91
Estimating Solutions, 93

    Lesson 25   Solving Word Problems with Fractions and Mixed Numbers, 94

    Lesson 26   Solving Multistep Word Problems with Fractions and Mixed Numbers, 99

Fractions Review   103

GED PRACTICE 1   105

UNIT 2   PERCENTS   109

Chapter 1   PERCENTS, DECIMALS, AND FRACTIONS   110

    Lesson 27   Understanding Percents, 110
Writing Percents, 110
Percents Larger than 100%, 113

    Lesson 28   Rewriting Percents as Decimals, 116
Moving the Decimal Point, 116
Using Zeros as Place Holders, 117
Rewriting Percents as Mixed Decimals, 118

    Lesson 29   Rewriting Decimals as Percents, 119
Moving the Decimal Point, 119
Rewriting Decimals with More than Two Places as Percents, 120
Rewriting Decimals with Only One Place as Percents, 121

Lesson 30   Rewriting Percents as Fractions, 122
  Rewriting Percents without Decimal Places as Fractions or Mixed Numbers, 123
  Rewriting Percents with Decimal Places as Fractions or Mixed Numbers, 124
  Rewriting Percents with Fractions as Fractions, 125

Lesson 31   Rewriting Fractions as Percents, 126
  Rewriting Fractions with Denominators of 100 as Percents, 126
  Rewriting Other Fractions as Percents, 127

Lesson 32   Recognizing Common Equivalents, 129

Chapter 2   SOLVING PERCENT PROBLEMS 131

Lesson 33   Understanding the Percent Triangle, 131
  Identifying the Numbers in a Percent Problem, 131
  Using the Percent Triangle, 132

Lesson 34   Finding the Part, 135
  Finding the Part by Using the Decimal Equivalent of a Percent, 135
  Finding the Part by Using the Fraction Equivalent of a Percent, 136

Lesson 35   Finding the Percent, 138
  Finding the Percent by Dividing, 138
  Finding the Percent by Simplifying a Fraction, 140
  Finding Percents Larger than 100%, 141

Lesson 36   Finding the Whole, 142
  Dividing by a Decimal to Find the Whole, 143
  Dividing by a Fraction to Find the Whole, 144

Percent Skills Review **146**

Chapter 3   WORKING WITH PERCENTS 148

Lesson 37   Estimating Solutions to Percent Problems, 148
  Estimating the Part by Changing the Whole, 148
  Estimating the Part by Changing the Percent, 150
  Estimating the Whole, 151
  Estimating the Percent, 152

Lesson 38   Solving One-Step Percent Word Problems, 153

Lesson 39   Solving Multistep Percent Problems, 158
  Finding Parts Larger than the Whole, 158
  Finding Parts Smaller than the Whole, 160
  Finding Percents of Change, 161

Percents Review **164**

GED PRACTICE 2 **167**

UNIT 3   TABLES AND GRAPHS   *171*

Chapter 1   UNDERSTANDING GRAPHIC INFORMATION   172

Lesson 40   Graphic Displays of Data, 172
Tables, 172
Bar Graphs, 176
Pictographs, 180

Lesson 41   Graphic Displays of Trends, 182
Line Graphs: Finding Information, 183
Line Graphs: Estimating, 184

Lesson 42   Graphic Displays of the Parts of Wholes, 185
Circle Graphs, 186
Divided-Bar Graphs, 188

Lesson 43   Double Graphs, 190
Double-Bar Graphs, 190
Double-Line Graphs, 193

Chapter 2   WORKING WITH GRAPHIC INFORMATION   197

Lesson 44   Solving One-Step Problems, 197
Adding to Find Facts, 197
Subtracting to Find Facts, 198
Multiplying to Find Facts, 201
Dividing to Find Facts, 201

Lesson 45   Solving Multistep Problems, 203

GED PRACTICE 3   *208*

POSTTEST   *211*

ANSWERS AND SOLUTIONS   *221*

ACKNOWLEDGMENTS

CAMBRIDGE Adult Education thanks the men and women enrolled in ABE and Pre-GED courses who read parts of the *Threshold* manuscripts and offered valuable advice to the programs' authors and editors.

We also thank the following consultants for their many contributions throughout the preparation of the *Threshold* Pre-GED programs.

Cecily Kramer Bodnar
Consultant, Adult Learning
Adult Literacy Services
Central School District
Greece, New York

Pamela S. Buchanan
Instructor
Blue Ridge Job Corps Center
Marion, Virginia

Maureen Considine, M.A., M.S.
Learning Laboratory Supervisor
Great Neck Adult Learning Center
Great Neck, New York

ABE/HSE Projects Coordinator
National Center for Disability Services
Albertson, New York

Carole Deletiner
Instructor
Hunter College
New York, New York

Patricia Giglio
Remedial Reading Teacher
Johnstown ASACTC
Johnstown, New York

Diane Marinelli Hardison, M.S. Ed.
Mathematics Educator
San Diego, California

Margaret Banker Tinzmann, Ph. D.
Program Associate
The North Central Regional Educational Laboratory
Oak Brook, Illinois

INTRODUCTION

The *Threshold* Pre-GED Programs

Threshold provides a full-range entry-level course for adults whose goal is to earn a high school equivalency diploma. The men and women who use the six *Threshold* programs will learn—and profit from an abundance of sound practice in applying—the writing, problem-solving, and critical-reading and -thinking skills they'll need when they take the GED tests. They will gain a firm grounding in knowledge about social studies and science and will read many excellent selections from the best of classical and contemporary literature. In short, *Threshold* offers adults the skills, knowledge, and practice that will enable them to approach GED-level test preparation with well-deserved confidence and solid ability.

The *Threshold* Mathematics Program

Two books make up the mathematics program. The first volume covers whole numbers and decimals, and this one covers fractions, percents, and tables and graphs.

Students should begin their study in this book by taking the Pretest. It has three parts, which can be administered at the same time or separately. To aid accurate assessment, the Pretest's problems are not multiple choice. To facilitate placement, the test has at least one problem related to each lesson in this book, as the Skills Chart that follows it shows.

The lessons present the various facets of fraction, percent, and graphic-material–related operations and applications in a carefully graded skill-building sequence. Lessons are typically divided into subskill segments, each with a succinct explanation, an example worked out and explained step by step, an exercise, and a word problem.

In Unit 1 Mixed Practices follow the chapters on subtraction and division. A comprehensive skills review precedes the last chapter in both Units 1 and 2.

Each unit's final chapters cover applications and/or one-step and multistep problem solving. The lessons show how to use the basic skills to solve the kinds of problems adults encounter at work, in their everyday lives, and on the GED.

Each of the three units ends with a GED Practice. The Practices, formatted like the GED, provide valuable test-taking experience. Accompanying skills charts allow assessment of skill mastery. (In Units 1 and 2 a thorough unit review precedes the Practice.)

The Posttest, also formatted like the GED, provides a comprehensive review of all the material covered in this book, as the Skills Chart that follows it shows.

This book's quick and accurate placement tool, its carefully segmented instruction, its exercises with word problems, its chapters on applications and problem solving, its progress-assessment charts, and its GED-like practices and Posttest make it—together with the first volume—an excellent first course in preparation for the mathematics test of the GED.

TO THE STUDENT

You will profit in several important ways by using this book as you begin to prepare for the mathematics test of the GED:
- You will improve your math skills.
- You will increase your problem-solving ability.
- You will gain experience in answering questions like those on the GED.
- You will become more confident of your abilities.

To Find Out About Your Current Math Skills . . .

Take the **PRETEST**. When you have finished, refer to the **ANSWERS AND SOLUTIONS** at the back of this book to check your answers. Then look at the **SKILLS CHART** that follows the Pretest. It will give you an idea about which parts of this book you need to concentrate on most.

To Improve Your Math Skills and Problem-Solving Ability . . .

Study the **LESSONS**. They present instruction about the various math skills and about the methods for solving problems. Each lesson includes one or more **EXAMPLES** and **EXERCISES** to help you improve both your math skills and your ability to solve problems.

Do the problems in the **MIXED PRACTICES** in Unit 1 and in the **REVIEWS** in Units 1 and 2. The problems they contain offer further practice and provide a way to review.

To Gain Experience in Solving Problems Like Those on the GED . . .

Take the **GED PRACTICE** at the end of each unit. The GED Practices are made up of problems like the ones on the mathematics test of the GED. They offer test-taking experience that you will find useful when you take the GED.

Before you finish with this book, take the **POSTTEST**. Like the three GED Practices, it is similar to the GED's mathematics test. Look at the **SKILLS CHART** that follows the Posttest. If you compare your Pretest and Posttest performances, you will probably find that your math skills and problem-solving ability have improved as you have worked through this book. The chart can give you an idea about which parts of this book you should review.

THRESHOLD

**Cambridge Pre-GED Program
in
Mathematics
2**

Pretest

If you begin this book at the very beginning, you may find that you already know how to do some of the work. This pretest can help you decide if you really need to start at page 10, or if you can skip some lessons. Answer as many of the questions as you can and then check your answers. The Skills Chart on page 7 will help you find where you should start your work in this book.

1

MATHEMATICS PRETEST

Directions: Answer as many questions as you can.

Fraction Problems

Item 1 refers to the following figure.

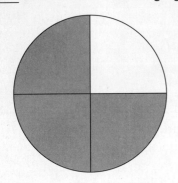

1. Write the fraction that represents the shaded part of this circle.

    _____

2. Simplify $\frac{6}{8}$.

3. Simplify $\frac{15}{60}$.

4. Raise $\frac{3}{4}$ to a fraction with 16 as its denominator.

5. Write these fractions with their lowest common denominator:

    $\frac{3}{4}$ $\frac{5}{6}$ $\frac{7}{12}$

6. Which of these fractions is the larger?

    $\frac{2}{3}$ $\frac{7}{10}$

7. Arrange the following fractions in order from smallest to largest.

    $\frac{3}{5}$ $\frac{7}{15}$ $\frac{2}{3}$ $\frac{4}{9}$

8. Change $\frac{29}{8}$ to a mixed number.

9. Change $2\frac{2}{5}$ to an improper fraction.

10. $\frac{11}{14} + \frac{9}{14} =$

11. $3\frac{5}{8} + 4\frac{5}{8} =$

12. $\frac{1}{5} + \frac{5}{6} + \frac{3}{10} =$

13. $3\frac{5}{12} + 6\frac{7}{8} =$

14. $\frac{13}{15} - \frac{8}{15} =$

15. $8\frac{3}{10} - 7\frac{1}{10} =$

16. $\frac{3}{4} - \frac{1}{3} =$

17. $13\frac{1}{4} - 5\frac{5}{6} =$

18. $\frac{1}{5} \times \frac{2}{3} =$

19. $\frac{3}{4} \times \frac{8}{9} =$

20. $9 \times \frac{3}{7} =$

21. $4\frac{4}{5} \times 2\frac{1}{2} =$

22. $1\frac{5}{6} \div \frac{1}{3} =$

23. $3\frac{3}{4} \div 3 =$

24. $3\frac{1}{3} \div 2\frac{2}{9} =$

25. 5 ft = _____ yd

26. $1\frac{1}{2}$ gal = _____ pt

Items 27 and 28 refer to the following rectangle.

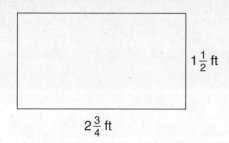

27. What is the perimeter of the rectangle?

28. What is the area of the rectangle?

Items 29 and 30 refer to the following triangle.

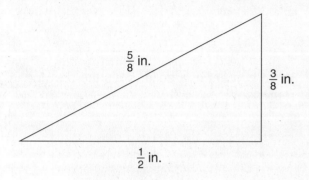

29. What is the perimeter of the triangle?

30. What is the area of the triangle?

Items 31 and 32 refer to the following circle.

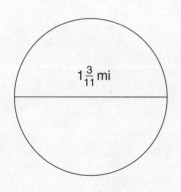

31. What is the circumference of the circle?

32. What is the area of the circle?

33. Change $\frac{7}{8}$ to a decimal.

34. Change .15 to a fraction.

35. Milagros bought a wedge of cheese that weighed 2.2 pounds. She used $\frac{5}{8}$ of it to make macaroni and cheese. How many pounds of the cheese did she use?

36. To estimate the sum of $2\frac{5}{8}$ and $14\frac{3}{7}$, round the two numbers and add. The estimated sum is
    (1) 16
    (2) 17
    (3) 18

37. Hermiña used equal portions of $6\frac{3}{8}$ cups of sugar in each of 3 batches of hard candy. How much did he use in each batch?

38. Gerta is hooking a thick rug for her breakfast nook, which measures $6\frac{3}{4}$ feet by 8 feet. The rug requires $1\frac{1}{6}$ pounds of wool yarn per square foot. How much yarn will Gerta need?

Pretest 3

Percent Problems

Item 39 refers to the following figure.

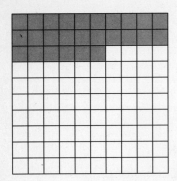

39. What percent of the figure is shaded?

Item 40 refers to the following drawing.

40. What percent of a dollar does the drawing represent?

41. Change 8% to a decimal.

42. Change .625 to a percent.

43. Change 87.5% to a fraction.

44. Change $\frac{3}{8}$ to a percent.

45. A sales brochure says that at 20% off you can save $80 on a $400 stereo. In the statement in the brochure,
    (a) which number is the whole?
    (b) which number is the percent?
    (c) which number is the part?

46. What is 35% of 200?

47. 18 is what percent of 40?

48. $82.50 is 15% of what amount?

49. Bill saved $33\frac{1}{3}$% on building materials that retailed for $2970. He saved about
    (a) $9000
    (b) $3000
    (c) $1000

50. Jerry paid 90% of the original price for carpeting that was $15 per square yard. How much did he pay per square yard?

51. Marty's business increased by 14% this year. Last year he did $15,000 in business. How much did he do this year?

Problems Based on Tables and Graphs

Items 52 through 55 refer to the following graph.

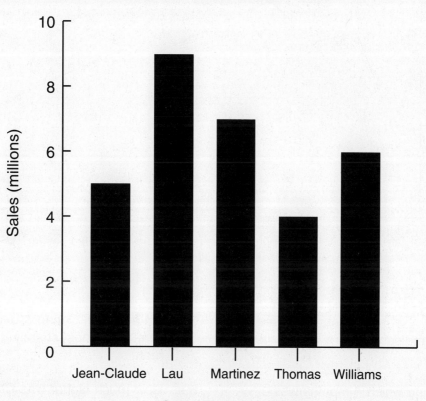

52. Which sales representative sold the most subscriptions to Worldwide Magazine in 1992?

53. How many subscriptions to Worldwide Magazine did Martinez sell in 1992?

54. In 1992, how many more subscriptions to Worldwide Magazine did Williams sell than Thomas?

55. Of all the Worldwide Magazine subscriptions sales in 1992, what percent were Lau's?

Items 56 through 59 refer to the following graph.

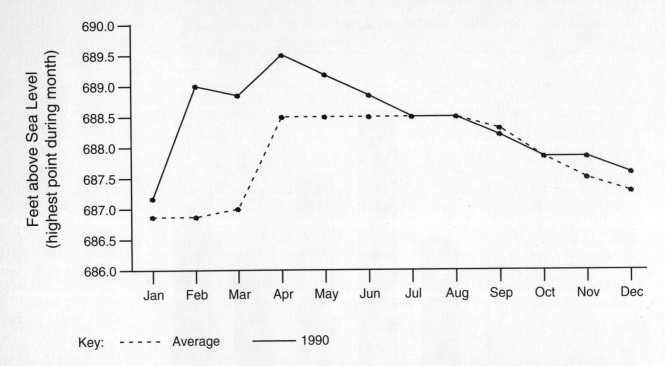

56. On average, how many feet above sea level is Canandaigua Lake's water level in June?

57. On average, Canandaigua Lake's water level rises most sharply between which two months?

58. In 1990, during which 4 months was Canandaigua Lake's water level the same as or lower than the average for those months?

59. How much higher was Canandaigua Lake's water level in April 1990 than in an average April?

Items 60 through 63 refer to the following graph.

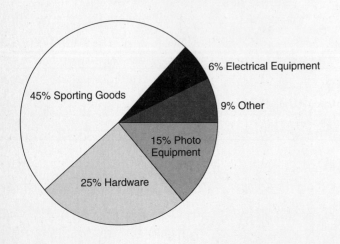

MILLER'S VARIETY STORE: 1992 SALES
TOTAL: $320,000

60. What type of sales accounted for the least business in Miller's Variety Store in 1992?

61. Hardware sales accounted for what percent of all sales in Miller's Variety Store in 1992?

62. What was the dollar value of hardware sales in Miller's Variety Store in 1992?

63. What was the dollar value of sales of photo equipment and sporting goods together in Miller's Variety Store in 1992?

Check your answers on page 221.

MATHEMATICS PRETEST SKILLS CHART

This chart shows which lesson covers the mathematics skills tested by each item in the Pretest. You can study as much of this book as you wish, but you should study at least those lessons related to any items you missed on the test.

Unit 1	Fractions	Item Number
Lesson 1	Proper Fractions	1
Lesson 2	Simplifying Proper Fractions	2, 3
Lesson 3	Raising Proper Fractions to Higher Terms	4
Lesson 4	Finding Lowest Common Denominators	5
Lesson 5	Comparing and Ordering Fractions	6, 7
Lesson 6	Improper Fractions and Mixed Numbers	8, 9
Lesson 7	Adding Fractions with Like Denominators	10
Lesson 8	Adding Mixed Numbers with Like Denominators	11
Lesson 9	Adding Fractions with Unlike Denominators	12
Lesson 10	Adding Mixed Numbers with Unlike Denominators	13
Lesson 11	Subtracting with Like Denominators	14, 15
Lesson 12	Subtracting with Unlike Denominators	16
Lesson 13	Borrowing	17
Lesson 14	Multiplying Fractions by Fractions	18
Lesson 15	Canceling before Multiplying Fractions	19
Lesson 16	Multiplying Whole Numbers by Fractions	20
Lesson 17	Multiplying Mixed Numbers by Fractions or Mixed Numbers	21
Lesson 18	Dividing by Fractions	22
Lesson 19	Dividing Fractions or Mixed Numbers by Whole Numbers	23
Lesson 20	Dividing by Mixed Numbers	24
Lesson 21	Working with Measurements Expressed as Fractions or Mixed Numbers	25, 26
Lesson 22	Finding Perimeters, Circumferences, and Areas	27, 28, 29, 30, 31, 32
Lesson 23	Working with Both Fractions and Decimals	33, 34, 35
Lesson 24	Rounding Mixed Numbers and Estimating Solutions	36
Lesson 25	Solving Word Problems with Fractions and Mixed Numbers	37
Lesson 26	Solving Multistep Word Problems with Fractions and Mixed Numbers	38

Unit 2   Percents

Lesson 27	Understanding Percents	39, 40
Lesson 28	Rewriting Percents as Decimals	41
Lesson 29	Rewriting Decimals as Percents	42
Lesson 30	Rewriting Percents as Fractions	43
Lesson 31	Rewriting Fractions as Percents	44
Lesson 32	Recognizing Common Equivalents	(42, 43, 44)
Lesson 33	Understanding the Percent Triangle	45
Lesson 34	Finding the Part	46
Lesson 35	Finding the Percent	47
Lesson 36	Finding the Whole	48
Lesson 37	Estimating Solutions to Percent Problems	49
Lesson 38	Solving One-Step Percent Word Problems	50
Lesson 39	Solving Multistep Percent Problems	51

Unit 3   Tables and Graphs

Lesson 40	Graphic Displays of Data	52, 53
Lesson 41	Graphic Displays of Trends	56, 57
Lesson 42	Graphic Displays of the Parts of Wholes	60, 61
Lesson 43	Double Graphs	58, 59
Lesson 44	Solving One-Step Problems	54, 62
Lesson 45	Solving Multistep Problems	55, 63

UNIT 1

Fractions

Chapter 1 of this unit introduces fractions. In Chapters 2–5, you will practice the four operations—addition, subtraction, multiplication, and division—using fractions. In Chapter 6 you will use fractions to solve everyday problems—problems like many of those on the GED. You will work with measurements; perimeter, circumference, and area; and a variety of one-step and multistep word problems.

Unit 1 Overview

Chapter 1   Understanding Fractions
Chapter 2   Addition
Chapter 3   Subtraction
Chapter 4   Multiplication
Chapter 5   Division
Chapter 6   Using Fractions

GED Practice 1

Chapter 1

UNDERSTANDING FRACTIONS

The six lessons in this chapter introduce fractions and mixed numbers. The lessons cover some of the skills you will need to use in later chapters where you will add, subtract, multiply, and divide fractions and mixed numbers.

In the first lesson you will learn what proper fractions represent. In the next two lessons you will simplify proper fractions and raise them to higher terms. In the fourth and fifth lessons you will find and use lowest common denominators. In the last lesson you will find out about improper fractions and mixed numbers. Then you can change improper fractions into mixed or whole numbers—or vice versa.

Lesson 1

Proper Fractions

A **fraction** stands for one or more parts of a whole amount. Fractions are made up of two numbers. The top number is the **numerator,** and the bottom number is the **denominator.**

$\frac{5}{6}$ ← The numerator tells the number of equal parts you have.
← The denominator tells how many equal parts make up the whole amount.

In a **proper fraction,** the numerator is smaller than the denominator. Proper fractions stand for amounts smaller than a whole—that is, smaller than 1.

This circle is divided into 4 parts. One part is shaded. The proper fraction $\frac{1}{4}$ stands for the shaded part. The proper fraction $\frac{3}{4}$ stands for the unshaded portion.

This circle is divided into 3 parts. Two parts are shaded. The proper fraction $\frac{2}{3}$ stands for the shaded portion. The proper fraction $\frac{1}{3}$ stands for the unshaded portion.

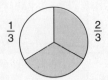

10   UNIT 1: Fractions

Writing Proper Fractions

To write a proper fraction, count how many parts make up the whole and write that number as the denominator. Then count how many parts you have and write that number as the numerator.

Example: Write a proper fraction that represents the shaded portion of this rectangle.

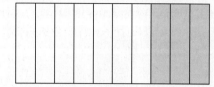

         Step 1   **Step 2**

          $\frac{}{10}$ $\frac{3}{10}$

STEP 1: Count the parts that make up the rectangle. There are 10. Write 10 as the denominator.

STEP 2: Count the shaded parts. There are 3. Write 3 as the numerator. The fraction is $\frac{3}{10}$.

EXERCISE 1

Part A. For each figure, write two proper fractions: (a) the fraction that represents the shaded portion and (b) the fraction that represents the unshaded portion.

1. (a) shaded portion _____

   (b) unshaded portion _____

2. (a) shaded portion _____

   (b) unshaded portion _____

3. (a) shaded portion _____

   (b) unshaded portion _____

4. (a) shaded portion _____

   (b) unshaded portion _____

5. (a) shaded portion _____
   (b) unshaded portion _____

6. (a) shaded portion _____
   (b) unshaded portion _____

 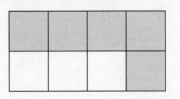

7. (a) shaded portion _____
   (b) unshaded portion _____

8. (a) shaded portion _____
   (b) unshaded portion _____

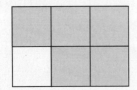

9. (a) shaded portion _____
   (b) unshaded portion _____

10. (a) shaded portion _____
    (b) unshaded portion _____

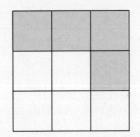 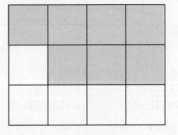

11. (a) shaded portion _____
    (b) unshaded portion _____

12. (a) shaded portion _____
    (b) unshaded portion _____

Part B. Shade the indicated portion of each figure.

1.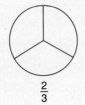
   $\frac{2}{3}$

2.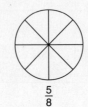
   $\frac{5}{8}$

3.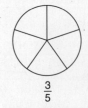
   $\frac{3}{5}$

12   UNIT 1: Fractions

4.
$\frac{4}{5}$

5.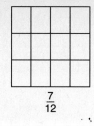
$\frac{7}{12}$

6.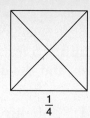
$\frac{1}{4}$

7.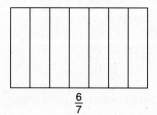
$\frac{6}{7}$

8.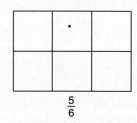
$\frac{5}{6}$

9.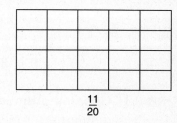
$\frac{11}{20}$

⎡ **WORD PROBLEM**

This problem asks you to write a proper fraction that stands for a portion of a work day.

Sandy works 8 hours a day. On Monday she spent 3 hours writing a report. Write a fraction that shows what portion of Monday Sandy spent on the report. ⎦

Check your answers on page 222.

Lesson 2

Simplifying Proper Fractions

Some proper fractions can be simplified, or written in **lower terms.** That is, they can be written with smaller numbers.

Compare these two figures. The same portion of each figure is shaded.

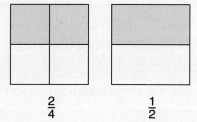

$\frac{2}{4}$ $\frac{1}{2}$

The fraction $\frac{2}{4}$ can be simplified to $\frac{1}{2}$. The value of $\frac{2}{4}$ does not change when it is simplified. It is just written in lower terms (that is, with smaller numbers). The fractions $\frac{2}{4}$ and $\frac{1}{2}$ are equal to each other. For that reason, they are called **equivalent fractions.**

Chapter 1: Understanding Fractions

Writing Fractions in Lower Terms

To **simplify** a fraction (to write it in lower terms), divide the numerator (top number) and the denominator (bottom number) by the same number.

Example 1: Simplify $\frac{6}{9}$.

Step 1	Step 2	Step 3
$\frac{6 \div 3}{9 \div 3}$	$\frac{6 \div 3}{9 \div 3} = \frac{2}{}$	$\frac{6 \div 3}{9 \div 3} = \frac{2}{3}$

STEP 1: Choose a number that can divide into both the numerator and the denominator evenly. 3 divides into both 6 and 9 evenly.

STEP 2: Divide 6 by 3. Write the result, 2, as the new numerator.

STEP 3: Divide 9 by 3. Write the result, 3, as the new denominator. The fraction $\frac{6}{9}$ has been simplified to $\frac{2}{3}$. The fraction $\frac{2}{3}$ is in lower terms than $\frac{6}{9}$ because the numbers used to write it are smaller.

EXERCISE 2a

Simplify each fraction.

1. $\frac{15}{21}$
2. $\frac{2}{24}$
3. $\frac{9}{15}$
4. $\frac{15}{20}$

5. $\frac{9}{18}$
6. $\frac{14}{35}$
7. $\frac{8}{10}$
8. $\frac{18}{27}$

9. $\frac{40}{100}$
10. $\frac{45}{100}$

> **WORD PROBLEM**
>
> This problem asks you to simplify a fraction.
>
> A baker used 6 pounds of flour from a 10-pound bag. In other words, he used $\frac{6}{10}$ of the flour. Express the portion of flour used in lower terms.

Check your answers on page 222.

UNIT 1: Fractions

Simplifying Fractions to Lowest Terms

A fraction has been simplified to its **lowest terms** when its numerator and denominator cannot be divided evenly by any number other than 1.

To write a fraction in its lowest terms, you may have to simplify more than once. That happens if you divide the numerator and denominator by too small a number on your first attempt.

Example 2: Simplify $\frac{16}{24}$.

Step 1	Step 2	Step 3	Step 4
$\frac{16 \div 4}{24 \div 4}$	$\frac{16 \div 4}{24 \div 4} = \frac{4}{6}$	$\frac{4 \div 2}{6 \div 2}$	$\frac{4 \div 2}{6 \div 2} = \frac{2}{3}$

STEP 1: Choose a number that can divide into both 16 and 24 evenly. 4 divides into both 16 and 24 evenly.

STEP 2: Using 4 to divide, simplify $\frac{16}{24}$. The answer is $\frac{4}{6}$.

STEP 3: Check to see if the answer is in lowest terms. It is not, because both the numerator and the denominator can be divided evenly by 2.

STEP 4: Using 2 to divide, simplify $\frac{4}{6}$. In lowest terms, the fraction $\frac{16}{24}$ is $\frac{2}{3}$.

Note: It is possible to simplify $\frac{16}{24}$ to $\frac{2}{3}$ in one step by dividing the numerator and denominator by 8. Try it.

EXERCISE 2b

Simplify each fraction. Be sure your answers are in lowest terms.

1. $\frac{8}{12}$
2. $\frac{10}{20}$
3. $\frac{6}{9}$
4. $\frac{20}{48}$

5. $\frac{12}{21}$
6. $\frac{28}{32}$
7. $\frac{3}{9}$
8. $\frac{12}{15}$

9. $\frac{6}{10}$
10. $\frac{9}{15}$
11. $\frac{9}{30}$
12. $\frac{7}{42}$

13. $\frac{30}{36}$
14. $\frac{10}{25}$
15. $\frac{4}{36}$
16. $\frac{4}{10}$

17. $\frac{20}{40}$
18. $\frac{24}{54}$
19. $\frac{16}{44}$
20. $\frac{18}{45}$

21. $\frac{35}{56}$ 22. $\frac{16}{18}$ 23. $\frac{19}{57}$ 24. $\frac{30}{55}$

25. $\frac{18}{72}$ 26. $\frac{8}{36}$ 27. $\frac{25}{75}$ 28. $\frac{16}{56}$

29. $\frac{14}{21}$ 30. $\frac{15}{100}$

WORD PROBLEM

The key phrase *what part* can be a clue that you need to write a fraction to solve a word problem.

To meet his quota, Kwok Eng must sell all the raffle tickets in a book of 24. He has sold 8 tickets. **What part** of the book has he sold? (Express the answer in lowest terms.)

Check your answers on page 223.

Lesson 3: Raising Proper Fractions to Higher Terms

Any fraction can be raised to **higher terms.** That is, it can be written as an equivalent fraction with larger numbers. Raising a fraction to higher terms is the opposite of simplifying it.

Compare these two figures. They represent partly filled bags of cement. Both bags have the same amount of cement in them.

$\frac{2}{3}$

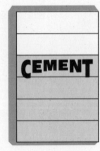
$\frac{4}{6}$

The fraction $\frac{2}{3}$ can be raised to $\frac{4}{6}$. The value of $\frac{2}{3}$ does not change when it is raised to higher terms. It is just written with larger numbers. The fractions $\frac{2}{3}$ and $\frac{4}{6}$ are equivalent fractions.

Multiplying by Any Number to Raise Terms

To raise a fraction to higher terms, multiply the numerator (the top number) and the denominator (the bottom number) by the same number—any number larger than 1.

Example 1: Raise $\frac{3}{5}$ to higher terms.

Step 1	Step 2	Step 3
$\frac{3 \times 2}{5 \times 2}$	$\frac{3 \times 2}{5 \times 2} = \frac{6}{\_}$	$\frac{3 \times 2}{5 \times 2} = \frac{6}{10}$

STEP 1: Choose a number to use to raise the fraction to higher terms. In this example, 2 is used.

STEP 2: Multiply 3 by 2. Write 6 as the new numerator.

STEP 3: Multiply 5 by 2. Write 10 as the new denominator. The fraction $\frac{3}{5}$ has been raised to $\frac{6}{10}$.

You may use as large a number as you wish to express a fraction in higher terms. If 3 had been used in the example above, the equivalent fraction would have been $\frac{9}{15}$. If 4 had been used, the equivalent fraction would have been $\frac{12}{20}$. If 20 had been used, the equivalent fraction would be $\frac{60}{100}$.

EXERCISE 3a

Part A. Raise each fraction to higher terms. Multiply both the numerator and denominator by the number that is given.

1. Raise $\frac{7}{8}$ to higher terms. Multiply each number by 2.

2. Raise $\frac{1}{4}$ to higher terms. Multiply each number by 3.

3. Raise $\frac{2}{3}$ to higher terms. Multiply each number by 4.

4. Raise $\frac{2}{5}$ to higher terms. Multiply each number by 5.

5. Raise $\frac{1}{10}$ to higher terms. Multiply each number by 6.

Part B. Change each fraction to three equivalent fractions with higher terms. In each case, choose your own number by which to multiply both numerator and denominator.

1. $\frac{1}{2}$ _____ _____ _____

2. $\frac{3}{4}$ _____ _____ _____

3. $\dfrac{5}{6}$ _____ _____ _____   4. $\dfrac{3}{8}$ _____ _____ _____

5. $\dfrac{5}{9}$ _____ _____ _____   6. $\dfrac{1}{3}$ _____ _____ _____

7. $\dfrac{1}{6}$ _____ _____ _____   8. $\dfrac{4}{5}$ _____ _____ _____

9. $\dfrac{5}{8}$ _____ _____ _____   10. $\dfrac{3}{10}$ _____ _____ _____

11. $\dfrac{2}{7}$ _____ _____ _____   12. $\dfrac{3}{12}$ _____ _____ _____

Check your answers on page 223.

Multiplying by a Certain Number to Raise Terms

Sometimes before you raise a fraction to higher terms, you will already know either the numerator or the denominator of the equivalent fraction. In those cases, you need to find what number to multiply by to get the other part of the equivalent fraction. In such problems, there is a division step.

Example 2: Raise $\dfrac{2}{3}$ to a fraction with 12 as its denominator.

Step 1	**Step 2**	**Step 3**
$\dfrac{2 \times ?}{3 \times ?} = \dfrac{?}{12}$	$\dfrac{2 \times 4}{3 \times 4} = \dfrac{?}{12}$	$\dfrac{2 \times 4}{3 \times 4} = \dfrac{8}{12}$

STEP 1: Set up the problem. Write in the numbers you know, and put question marks for the numbers you don't know.

STEP 2: Ask yourself what number multiplied by 3 gives 12. To get the answer, divide: $12 \div 3 = 4$. Write 4 in both the numerator and denominator of the fraction you are raising.

STEP 3: Multiply 2 by 4. Write 8 as the new numerator. The fraction $\dfrac{2}{3}$ has been raised to $\dfrac{8}{12}$.

The example above shows how to raise a fraction to higher terms when you know only the new denominator. The same method also works when you know only the new numerator.

EXERCISE 3b

Part A. Raise each fraction to higher terms by writing the numerator for the equivalent fraction.

1. $\dfrac{1}{5} = \dfrac{?}{20}$
2. $\dfrac{1}{6} = \dfrac{?}{36}$
3. $\dfrac{1}{10} = \dfrac{?}{30}$

4. $\dfrac{1}{3} = \dfrac{?}{18}$
5. $\dfrac{1}{5} = \dfrac{?}{45}$
6. $\dfrac{5}{8} = \dfrac{?}{24}$

7. $\dfrac{1}{3} = \dfrac{?}{9}$
8. $\dfrac{2}{3} = \dfrac{?}{9}$
9. $\dfrac{3}{4} = \dfrac{?}{8}$

10. $\dfrac{7}{10} = \dfrac{?}{100}$

Part B. Raise each fraction to higher terms by writing the denominator for the equivalent fraction.

1. $\dfrac{2}{5} = \dfrac{4}{?}$
2. $\dfrac{1}{12} = \dfrac{3}{?}$
3. $\dfrac{7}{10} = \dfrac{28}{?}$

4. $\dfrac{3}{7} = \dfrac{6}{?}$
5. $\dfrac{4}{5} = \dfrac{24}{?}$
6. $\dfrac{11}{16} = \dfrac{22}{?}$

7. $\dfrac{3}{7} = \dfrac{15}{?}$
8. $\dfrac{8}{9} = \dfrac{32}{?}$
9. $\dfrac{3}{5} = \dfrac{6}{?}$

10. $\dfrac{3}{8} = \dfrac{9}{?}$

WORD PROBLEM

This problem asks you to find the numerator for a fraction raised to higher terms.

Hannah wants to pump $\frac{1}{2}$ gallon of gasoline into the can she uses to fill her lawn mower. At the gas station, the pump measures gasoline in tenths of a gallon. Express the amount of gasoline Hannah needs as a fraction with 10 as the denominator.

Check your answers on page 223.

Chapter 1: Understanding Fractions

Lesson 4

Finding Lowest Common Denominators

To solve many problems, you need to rewrite all the fractions in the problem with the same denominator, that is, with a *common* denominator. In most such problems, it is best to use the **lowest common denominator,** or LCD. The **LCD** is the smallest denominator you can use to rewrite all the fractions so that they have the same denominator.

Using Multiples to Find the LCD

To rewrite two fractions using their LCD, first list some of the multiples of each denominator. Find the lowest multiple that is common to both denominators. Then, raise the terms of one or both of the fractions so that they have the same denominator, the LCD.

Example 1: Rewrite $\frac{3}{8}$ and $\frac{7}{12}$ using their LCD.

Step 1

$8 \times 1 = 8 \quad 12 \times 1 = 12$
$8 \times 2 = 16 \quad 12 \times 2 = \mathbf{24}$
$8 \times 3 = \mathbf{24} \quad 12 \times 3 = 36$
$8 \times 4 = 32 \quad 12 \times 4 = 48$
$8 \times 5 = 40 \quad$ Etc.
$8 \times 6 = 48$
Etc.

Step 2

$\frac{3 \times 3}{8 \times 3} = \frac{9}{24}$

$\frac{7 \times 2}{12 \times 2} = \frac{14}{24}$

Step 3

$\frac{9}{24}$ and $\frac{14}{24}$

STEP 1: List some of the multiples of the denominators 8 and 12. The lowest multiple that is common for 8 and 12 is 24. (Notice that 48 is also a common multiple for 8 and 12, but it is not the lowest one.)

STEP 2: Raise each fraction to an equivalent fraction with 24 as the denominator. (You learned to raise fractions to higher terms in Lesson 3.)

STEP 3: Write the two fractions using their LCD, 24.

You can use the same procedure to rewrite more than two fractions using their LCD. Suppose you want to rewrite $\frac{1}{5}, \frac{1}{3}$ and $\frac{1}{2}$ using their LCD. Some of the multiples of the denominators are as follows:

Multiples of 5: 10, 15, 20, 25, **30,** 35 . . .

Multiples of 3: 6, 9, 12, 15, 18, 21, 24, 27, **30,** 33 . . .

Multiples of 2: 4, 6, 8, 10, 12, 14, 16, 18, 20, 22, 24, 26, 28, **30,** 32 . . .

The lowest multiple common to all three is 30, so you would rewrite the fractions using 30 as the LCD: $\frac{6}{30}, \frac{10}{30},$ and $\frac{15}{30}$.

EXERCISE 4a

Rewrite the fractions in each set using their LCD.

1. $\frac{1}{6}$ and $\frac{3}{8}$
2. $\frac{1}{4}$ and $\frac{7}{10}$
3. $\frac{1}{8}$ and $\frac{1}{10}$
4. $\frac{1}{10}$ and $\frac{1}{12}$
5. $\frac{1}{4}$ and $\frac{5}{6}$
6. $\frac{5}{6}$ and $\frac{4}{15}$
7. $\frac{1}{2}, \frac{1}{3},$ and $\frac{1}{9}$
8. $\frac{1}{2}, \frac{2}{3},$ and $\frac{1}{4}$
9. $\frac{1}{4}, \frac{2}{3},$ and $\frac{1}{6}$
10. $\frac{5}{8}, \frac{3}{10},$ and $\frac{3}{4}$
11. $\frac{2}{3}, \frac{1}{9},$ and $\frac{3}{12}$
12. $\frac{4}{5}, \frac{1}{20},$ and $\frac{6}{25}$

Check your answers on page 223.

Dividing Denominators to Find the LCD

Sometimes you can use a kind of shortcut to find the LCD for two fractions. When the smaller denominator divides evenly into the larger denominator, the larger denominator is the LCD.

Example 2: Rewrite $\frac{2}{3}$ and $\frac{1}{6}$ using their LCD.

Step 1	Step 2	Step 3
$3\overline{)6}^{\,2}$	$\frac{2 \times 2}{3 \times 2} = \frac{4}{6}$	$\frac{4}{6}$ and $\frac{1}{6}$

STEP 1: The smaller denominator divides into the larger denominator evenly: $6 \div 3 = 2$. So 6 is the LCD.

STEP 2: Raise $\frac{2}{3}$ to an equivalent fraction with 6 as the denominator.

STEP 3: Rewrite the two fractions using their LCD, 6.

This shortcut often works when you want to find the LCD for more than two fractions. For example, suppose you want to find the LCD for $\frac{1}{2}, \frac{1}{4}$ and $\frac{1}{8}$. Both 2 and 4 divide into 8 evenly, so 8 is the LCD.

EXERCISE 4b

Rewrite the fractions in each set using their LCD.

1. $\frac{1}{2}$ and $\frac{1}{8}$
2. $\frac{3}{5}$ and $\frac{3}{10}$
3. $\frac{1}{12}$ and $\frac{5}{24}$
4. $\frac{4}{11}$ and $\frac{5}{22}$
5. $\frac{5}{9}$ and $\frac{1}{3}$
6. $\frac{2}{5}$ and $\frac{4}{15}$
7. $\frac{3}{8}, \frac{5}{16}$, and $\frac{1}{4}$
8. $\frac{1}{2}, \frac{1}{5}$, and $\frac{3}{20}$
9. $\frac{1}{6}, \frac{1}{9}$, and $\frac{5}{36}$
10. $\frac{1}{3}, \frac{1}{6}$, and $\frac{1}{12}$
11. $\frac{1}{5}, \frac{1}{15}$, and $\frac{11}{30}$
12. $\frac{1}{3}, \frac{8}{9}$, and $\frac{11}{27}$

Check your answers on page 223.

Multiplying Denominators to Find the LCD

There is another shortcut for finding the LCD for two fractions. Sometimes the LCD is the product of both denominators in a problem.

Example 3: Write $\frac{1}{3}$ and $\frac{3}{5}$ with their LCD.

Step 1   **Step 2**   **Step 3**

$3 \times 5 = 15$ $\frac{1 \times 5}{3 \times 5} = \frac{5}{15}$ $\frac{5}{15}$ and $\frac{9}{15}$

$\frac{3 \times 3}{5 \times 3} = \frac{9}{15}$

STEP 1: To find the LCD, multiply the denominator of the two fractions: $3 \times 5 = 15$.

STEP 2: Raise each fraction to an equivalent fraction with 15 as the denominator.

STEP 3: Write the two fractions with their LCD, 15.

You should use this shortcut only when there is no number, such as 2, that divides evenly into the denominators. In Example 3, there is no number except 1 that divides evenly into both 3 and 5. Thus the LCD is 3×5.

UNIT 1: Fractions

If you use this shortcut and the product of two denominators seems too large to be the LCD, list the multiples of the denominators. Then you can see if a smaller multiple is the LCD.

Multiplying denominators to find the LCD sometimes works when there are more than two fractions. To find the LCD for $\frac{1}{5}$, $\frac{1}{4}$ and $\frac{1}{3}$, you can multiply all three denominators together: $5 \times 4 \times 3 = 60$. Notice there is no number except 1 that divides evenly into all the denominators.

EXERCISE 4c

Rewrite the fractions in each set using their LCD.

1. $\frac{1}{4}$ and $\frac{4}{15}$
2. $\frac{1}{3}$ and $\frac{1}{2}$
3. $\frac{3}{4}$ and $\frac{1}{3}$

4. $\frac{4}{7}$ and $\frac{2}{3}$
5. $\frac{4}{5}$ and $\frac{3}{4}$
6. $\frac{1}{2}$ and $\frac{4}{5}$

7. $\frac{1}{2}, \frac{1}{3}$, and $\frac{1}{7}$
8. $\frac{1}{3}, \frac{1}{4}$, and $\frac{1}{7}$
9. $\frac{1}{4}, \frac{1}{3}$, and $\frac{1}{5}$

10. $\frac{1}{2}, \frac{2}{3}$, and $\frac{1}{5}$
11. $\frac{1}{3}, \frac{2}{5}$, and $\frac{1}{8}$
12. $\frac{1}{2}, \frac{3}{5}$, and $\frac{7}{9}$

Check your answers on page 223.

Lesson 5
Comparing and Ordering Fractions

When fractions have the same denominator, it is easy to compare them to find out which is the larger. The fraction with the larger numerator stands for the larger amount. For example, $\frac{5}{7}$ is larger than $\frac{2}{7}$.

Likewise, when several fractions have the same denominator, it is easy to write them in a list from smallest to largest. The fraction with the smallest numerator is the smallest; the one with the largest numerator is the largest. The fractions $\frac{5}{11}, \frac{3}{11}$, and $\frac{8}{11}$ arranged from smallest to largest are $\frac{3}{11}, \frac{5}{11}$, and $\frac{8}{11}$.

Chapter 1: Understanding Fractions 23

Comparing Fractions with Unlike Denominators

Fractions can be compared only when they have the same denominator.

The fractions $\frac{2}{5}$ and $\frac{3}{7}$ have different denominators. To compare them, raise them to equivalent fractions using their LCD as the denominator. Because $\frac{15}{35}$ is larger than $\frac{14}{35}$, $\frac{3}{7}$ is larger than $\frac{2}{5}$.

$$\frac{2 \times 7}{5 \times 7} = \frac{14}{35} \qquad \frac{3 \times 5}{7 \times 5} = \frac{15}{35}$$

To compare two fractions with different denominators, rewrite the fractions using their LCD. (You learned to rewrite fractions using their LCDs in Lesson 4.) Then compare the numerators of the new fractions.

Example 1: Which is larger, $\frac{2}{3}$ or $\frac{3}{4}$?

Step 1	**Step 2**	**Step 3**
$3 \times 4 = 12$	$\frac{2 \times 4}{3 \times 4} = \frac{8}{12}$	$\frac{3}{4}$ is larger than $\frac{2}{3}$
	$\frac{3 \times 3}{4 \times 3} = \frac{9}{12}$	

STEP 1: Find the LCD for the two fractions. In this case, it is easiest to multiply the denominators: $3 \times 4 = 12$.

STEP 2: Raise each fraction to an equivalent fraction using 12 as the denominator. Compare the numerators: 9 is larger than 8.

STEP 3: Because $\frac{9}{12}$ is larger than $\frac{8}{12}$, the larger fraction is $\frac{3}{4}$.

EXERCISE 5a

Find the larger fraction in each pair.

1. $\frac{7}{10}$ or $\frac{11}{15}$
2. $\frac{5}{7}$ or $\frac{2}{3}$
3. $\frac{3}{8}$ or $\frac{7}{12}$
4. $\frac{7}{8}$ or $\frac{2}{3}$
5. $\frac{9}{12}$ or $\frac{5}{9}$
6. $\frac{1}{4}$ or $\frac{2}{3}$

7. $\frac{3}{5}$ or $\frac{2}{3}$ 8. $\frac{1}{4}$ or $\frac{3}{16}$ 9. $\frac{3}{4}$ or $\frac{5}{6}$

10. $\frac{2}{3}$ or $\frac{3}{7}$ 11. $\frac{5}{9}$ or $\frac{6}{7}$ 12. $\frac{1}{7}$ or $\frac{2}{11}$

13. $\frac{1}{3}$ or $\frac{4}{9}$ 14. $\frac{1}{10}$ or $\frac{1}{6}$ 15. $\frac{1}{3}$ or $\frac{4}{15}$

> **WORD PROBLEM**
>
> Package A holds $\frac{13}{16}$ pound of rice. Package B holds $\frac{7}{8}$ pound of rice. Which package holds more rice?

Check your answers on page 224.

Ordering Fractions with Unlike Denominators

When you order fractions, you write them in a list from smallest to largest, or vice versa. To do that, it is necessary to compare the fractions. If their denominators are not the same, rewrite the fractions using their LCD.

Example 2: Arrange $\frac{1}{2}, \frac{3}{8}, \frac{5}{16}$, and $\frac{9}{32}$ in order from smallest to largest.

Step 1

$32 \div 2 = 16$
$32 \div 8 = 4$
$32 \div 16 = 2$

Step 2

$\frac{1 \times 16}{2 \times 16} = \frac{16}{32}$

$\frac{3 \times 4}{8 \times 4} = \frac{12}{32}$

$\frac{5 \times 2}{16 \times 2} = \frac{10}{32}$

$\frac{9}{32} = \frac{9}{32}$

Step 3

$\frac{9}{32}, \frac{5}{16}, \frac{3}{8}, \frac{1}{2}$

STEP 1: Find the LCD for the fractions. In this case, it is 32, the largest denominator, because all the smaller denominators divide into it evenly.

STEP 2: Rewrite the fractions so that each has 32 as its denominator. Compare the numerators. Because the smallest numerator is 9, the fraction $\frac{9}{32}$ is the smallest. Because the largest numerator is 16, the fraction $\frac{1}{2}$ is the largest.

STEP 3: Write the four original fractions in order from smallest to largest.

Chapter 1: Understanding Fractions

EXERCISE 5b

Arrange the fractions in each set in order from smallest to largest.

1. $\frac{1}{3}, \frac{1}{2}, \frac{1}{4}$
2. $\frac{1}{2}, \frac{1}{5}, \frac{1}{10}$
3. $\frac{3}{8}, \frac{3}{4}, \frac{5}{16}$
4. $\frac{3}{4}, \frac{2}{3}, \frac{7}{24}$
5. $\frac{1}{2}, \frac{7}{8}, \frac{5}{6}, \frac{2}{3}$
6. $\frac{1}{4}, \frac{7}{12}, \frac{2}{3}, \frac{5}{12}$
7. $\frac{1}{6}, \frac{1}{8}, \frac{1}{4}$
8. $\frac{1}{3}, \frac{7}{15}, \frac{4}{5}$
9. $\frac{5}{9}, \frac{2}{5}, \frac{7}{15}$
10. $\frac{2}{3}, \frac{3}{8}, \frac{5}{16}, \frac{7}{12}$
11. $\frac{5}{6}, \frac{5}{9}, \frac{3}{4}, \frac{2}{3}$
12. $\frac{5}{14}, \frac{3}{7}, \frac{1}{2}, \frac{6}{7}$

WORD PROBLEM

The weights of four coins are as follows: Coin A, $\frac{7}{10}$ ounce; Coin B, $\frac{3}{4}$ ounce; Coin C, $\frac{3}{5}$ ounce; and Coin D, $\frac{1}{2}$ ounce. List the coins in order of size from smallest to largest.

Check your answers on page 224.

Lesson 6

Improper Fractions and Mixed Numbers

In Lesson 1 you learned what a proper fraction is. In a proper fraction, the numerator is smaller than the denominator. A proper fraction stands for an amount smaller than 1.

This lesson is about the other two kinds of fractions: improper fractions and mixed numbers.

Writing Improper Fractions

In an **improper fraction,** the numerator is equal to or larger than the denominator. An improper fraction can stand for 1 or for an amount larger than 1.

26 UNIT 1: Fractions

This circle is divided into 4 parts. All four parts are shaded. The improper fraction $\frac{4}{4}$ stands for the shaded part—the whole circle, or 1 circle.

$\frac{4}{4}$

These circles are each divided into 3 parts. Five parts are shaded. The improper fraction $\frac{5}{3}$ stands for the shaded portion of **both** circles.

$\frac{5}{3}$

To write an improper fraction, count how many parts make up *one* whole and write that number as the denominator. Then count how many parts you have and write that number as the numerator.

Example 1: Write an improper fraction that represents the shaded portion of both of these squares.

     Step 1   **Step 2**

      $\frac{}{4}$ $\frac{7}{4}$

STEP 1: Count the parts that make up *one* square. There are 4. Write 4 as the denominator.

STEP 2: Count the shaded parts in both squares. There are 7. Write 7 as the numerator. The fraction is $\frac{7}{4}$.

The fraction $\frac{7}{4}$ is improper. Its numerator is larger than its denominator, and it stands for an amount larger than 1.

EXERCISE 6a

Part A. Tell whether each of the following fractions is proper or improper.

1. $\frac{3}{4}$ 2. $\frac{4}{3}$ 3. $\frac{23}{17}$ 4. $\frac{36}{40}$ 5. $\frac{9}{5}$

6. $\frac{4}{5}$ 7. $\frac{18}{15}$ 8. $\frac{8}{9}$ 9. $\frac{6}{6}$ 10. $\frac{6}{7}$

Chapter 1: Understanding Fractions

11. $\frac{7}{6}$ 12. $\frac{7}{7}$ 13. $\frac{12}{2}$ 14. $\frac{5}{4}$ 15. $\frac{8}{5}$

16. $\frac{5}{7}$ 17. $\frac{3}{3}$ 18. $\frac{3}{2}$ 19. $\frac{2}{3}$ 20. $\frac{1}{1}$

Part B. Write an improper fraction that represents the shaded portion of all the figures in each of the following problems.

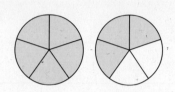

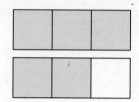

1. _____

2. _____

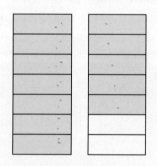

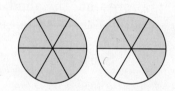

3. _____

4. _____

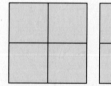

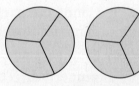

5. _____

6. _____

WORD PROBLEM

One Sunday Ralph baked two tins of muffins. Each tin held 8 muffins. At breakfast, his family ate all the muffins from one tin and three of the muffins from the other tin. Write an improper fraction that shows what portion of both tins of muffins the family ate.

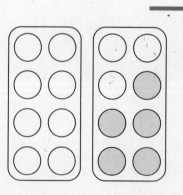

The unshaded circles stand for the muffins that were eaten.

Check your answers on page 224.

28   UNIT 1: Fractions

Writing Mixed Numbers

A **mixed number** is made up of a whole number and a fraction. A mixed number stands for an amount larger than 1.

In this figure, one whole circle and part of another are shaded. The mixed number $1\frac{2}{5}$ stands for the shaded portion of the two circles.

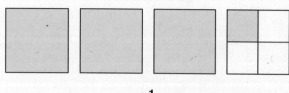

In this figure, three whole squares and part of another are shaded. The mixed number $3\frac{1}{4}$ stands for the shaded portion of the four squares.

To write a mixed number, count how many whole objects you have and write that amount as a whole number. Then write a proper fraction that stands for the remaining partial object.

Example 2: Write a mixed number that represents the shaded portion of all these circles.

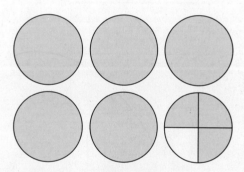

Step 1   **Step 2**

5        $5\frac{3}{4}$

STEP 1: Count how many whole circles there are. There are 5. Write 5 as a whole number.

STEP 2: Write a proper fraction that shows how much of the sixth circle is shaded. The circle has four parts, three of which are shaded. The fraction is $\frac{3}{4}$.

EXERCISE 6b

Write a mixed number that represents the shaded portion of all the figures in each of the following problems.

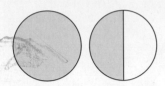

1. _____

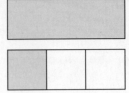

2. _____

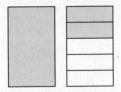

3. _____

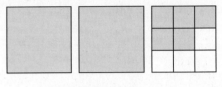

4. _____

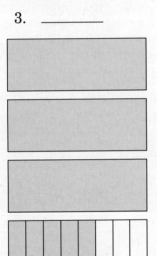

5. _____

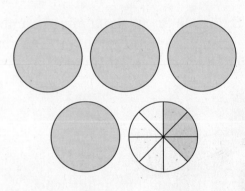

6. _____

WORD PROBLEM

On another Sunday, Ralph again baked two tins of 8 muffins. This time, his family ate one whole tin of muffins and five of the muffins from the other tin. Write a mixed number that shows how many tins of muffins the family ate.

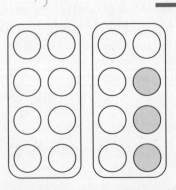

The unshaded circles stand for the muffins that were eaten.

Check your answers on page 224.

Rewriting Improper Fractions as Mixed or Whole Numbers

To rewrite an improper fraction as a mixed number or a whole number, divide the numerator by the denominator.

Example 3: Rewrite $\frac{20}{6}$ as a mixed number.

Step 1

$$6\overline{)20}^{\,3\ r2}\underline{18}2$$

Step 2

$$6\overline{)20}^{\,3\frac{2}{6}}\underline{18}2$$

Step 3

$$3\frac{2\div 2}{6\div 2} = 3\frac{1}{3}$$

STEP 1: Divide 20 by 6. The result is 3 with a remainder of 2.

STEP 2: Write the remainder as the numerator in a fraction with 6, the number you divided by, as the denominator.

STEP 3: Simplify the fraction. The improper fraction $\frac{20}{6}$ equals the mixed number $3\frac{1}{3}$.

If you divide to rewrite an improper fraction and there is no remainder, the answer will be a whole number. For example, when you divide to change $\frac{4}{4}$, the answer is 1. The improper fraction $\frac{12}{3}$ equals 4.

EXERCISE 6c

Rewrite each of these improper fractions as a mixed number or a whole number.

1. $\frac{3}{2}$ = $1\frac{1}{2}$
2. $\frac{21}{9}$ = $2\frac{3}{9} = 2\frac{1}{3}$
3. $\frac{7}{7}$ = 1
4. $\frac{4}{3}$ = $1\frac{1}{3}$
5. $\frac{7}{4}$ = $1\frac{3}{4}$
6. $\frac{35}{10}$ = $3\frac{5}{10} = 3\frac{1}{2}$
7. $\frac{10}{7}$ = $1\frac{3}{7}$
8. $\frac{18}{2}$ = 9
9. $\frac{14}{8}$ = $1\frac{6}{8} = 1\frac{3}{4}$
10. $\frac{15}{8}$ = $1\frac{7}{8}$
11. $\frac{5}{4}$ = $1\frac{1}{4}$
12. $\frac{8}{4}$ = 2
13. $\frac{14}{5}$ = $2\frac{4}{5}$
14. $\frac{13}{10}$ = $1\frac{3}{10}$
15. $\frac{11}{8}$ = $1\frac{3}{8}$
16. $\frac{4}{1}$ = 4
17. $\frac{7}{6}$ = $1\frac{1}{6}$
18. $\frac{9}{3}$ = 3
19. $\frac{12}{6}$ = 2
20. $\frac{8}{8}$ = 1

> **WORD PROBLEM**
>
> Lena ordered three pizzas for her guests. Each pizza was cut into 8 pieces. Lena and her guests ate $\frac{20}{8}$ of the pizzas. Express the amount they ate as a mixed number.

Check your answers on page 224.

Rewriting Mixed Numbers as Improper Fractions

To rewrite a mixed number as an improper fraction, multiply the whole number by the denominator. Then, add the numerator to the product. Finally, write the sum over the denominator of the fraction.

Example 4: Rewrite $2\frac{3}{4}$ as an improper fraction.

Note: The arrows and signs in the following illustration may help you to follow the steps more easily.

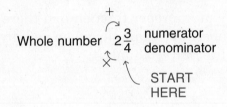

Step 1	Step 2	Step 3
$2 \times 4 = 8$	$8 + 3 = 11$	$\frac{11}{4}$

STEP 1: Multiply 2 (the whole number) by 4 (the denominator). The product is 8.

STEP 2: Add 3 (the numerator) to 8 (the product). The sum is 11.

STEP 3: Write 11 (the sum) over 4 (the denominator). The mixed number $2\frac{3}{4}$ equals the improper fraction $\frac{11}{4}$.

EXERCISE 6d

Rewrite each of these mixed numbers as improper fractions.

1. $2\frac{1}{2}$
2. $6\frac{4}{5}$
3. $3\frac{1}{3}$
4. $4\frac{1}{4}$

5. $12\frac{1}{2}$
6. $4\frac{1}{9}$
7. $5\frac{4}{5}$
8. $10\frac{2}{3}$

9. $8\frac{5}{7}$ 10. $10\frac{3}{5}$ 11. $4\frac{1}{10}$ 12. $3\frac{1}{5}$

13. $5\frac{7}{11}$ 14. $4\frac{5}{16}$ 15. $9\frac{1}{8}$ 16. $5\frac{5}{6}$

17. $7\frac{3}{4}$ 18. $5\frac{4}{5}$ 19. $1\frac{2}{3}$ 20. $9\frac{5}{6}$

WORD PROBLEM

The day Archie's car broke down, he walked $2\frac{3}{8}$ miles to work. Express the distance Archie walked as an improper fraction.

Check your answers on page 224.

Chapter 2 ADDITION

The four lessons in this chapter cover the addition of fractions and mixed numbers. The first two lessons cover all the skills you need in order to add fractions with the same denominator. The second two lessons cover all the skills you need in order to add fractions with different denominators.

Lesson 7: Adding Fractions with Like Denominators

To add fractions that have the same denominator, add their numerators and write the total over the denominator. Simplify the sum, if possible.

Example: Add $\frac{1}{10} + \frac{3}{10}$.

$$\frac{1}{10} \quad + \quad \frac{3}{10}$$

Step 1

Horizontal Method: $\frac{1}{10} + \frac{3}{10} =$

Vertical Method:
$$\frac{1}{10}$$
$$+ \frac{3}{10}$$

Step 2

$\frac{1}{10} + \frac{3}{10} = \frac{4}{10}$

$$\frac{1}{10}$$
$$+ \frac{3}{10}$$
$$\overline{\frac{4}{10}}$$

Step 3

$\frac{4 \div 2}{10 \div 2} = \frac{2}{5}$

$\frac{4 \div 2}{10 \div 2} = \frac{2}{5}$

STEP 1: Set up the problem. (Use the horizontal or the vertical method, whichever you prefer.)

STEP 2: Add the numerators: 1 + 3 = 4. Write 4 over 10, the denominator. The sum is $\frac{4}{10}$.

STEP 3: Divide 2 into both numerator and denominator to simplify the sum. The sum is simplified to $\frac{2}{5}$.

When a sum is an improper fraction, rewrite it as a mixed number or a whole number. Remember to simplify the fraction in each mixed number.

EXERCISE 7

Find each sum.

1. $\dfrac{1}{4} + \dfrac{2}{4}$

2. $\dfrac{7}{15} + \dfrac{5}{15}$

3. $\dfrac{6}{10} + \dfrac{3}{10}$

4. $\dfrac{1}{6} + \dfrac{3}{6} =$

5. $\dfrac{5}{12} + \dfrac{1}{12}$

6. $\dfrac{3}{7} + \dfrac{2}{7}$

7. $\dfrac{5}{9} + \dfrac{2}{9}$

8. $\dfrac{2}{5} + \dfrac{2}{5} =$

9. $\dfrac{17}{12} + \dfrac{3}{12}$

10. $\dfrac{7}{100} + \dfrac{13}{100}$

11. $\dfrac{3}{8} + \dfrac{7}{8}$

12. $\dfrac{1}{6} + \dfrac{5}{6} =$

13. $\dfrac{5}{8} + \dfrac{1}{8}$

14. $\dfrac{2}{5} + \dfrac{1}{5}$

15. $\dfrac{1}{6} + \dfrac{4}{6}$

16. $\dfrac{7}{12} + \dfrac{5}{12} =$

17. $\dfrac{8}{9} + \dfrac{8}{9}$

18. $\dfrac{3}{4} + \dfrac{1}{4}$

19. $\dfrac{11}{16} + \dfrac{2}{16}$

20. $\dfrac{7}{9} + \dfrac{4}{9} =$

21. $\dfrac{1}{9} + \dfrac{2}{9} + \dfrac{3}{9}$

22. $\dfrac{3}{5} + \dfrac{2}{5} + \dfrac{4}{5}$

23. $\dfrac{7}{12} + \dfrac{5}{12} + \dfrac{1}{12}$

24. $\dfrac{8}{15} + \dfrac{2}{15} + \dfrac{4}{15} =$

WORD PROBLEM

In the following problem, the key word *altogether* tells you to add.

Juan spent $\frac{1}{8}$ of his workday filing invoices. He spent $\frac{3}{8}$ of his workday planning a business trip. **Altogether,** what part of his workday did Juan spend filing and planning?

Check your answers on page 225.

Adding Mixed Numbers with Like Denominators

To add mixed numbers whose fractions have the same denominator, first add the whole numbers and then add the fractions.

Simplifying Sums with Fractions Smaller than One

Often the fraction in the sum of two or more mixed numbers is a proper fraction. It stands for an amount smaller than one. Simplify such a fraction, if possible.

Example 1: Add $4\frac{5}{8} + 3\frac{1}{8}$.

Step 1	Step 2	Step 3	Step 4
$4\frac{5}{8}$	$4\frac{5}{8}$	$4\frac{5}{8}$	$7\frac{6 \div 2}{8 \div 2} = 7\frac{3}{4}$
$+\ 3\frac{1}{8}$	$+\ 3\frac{1}{8}$	$+\ 3\frac{1}{8}$	
	7	$7\frac{6}{8}$	

STEP 1: Set up the problem. (Most people prefer to use the vertical method with mixed numbers.)

STEP 2: Add the whole numbers: $4 + 3 = 7$.

STEP 3: Add the fractions: $\frac{5}{8} + \frac{1}{8} = \frac{6}{8}$.

STEP 4: The sum is $7\frac{6}{8}$. Divide the numerator and denominator of the fraction by 2 to simplify it. The sum is $7\frac{3}{4}$.

36 UNIT 1: Fractions

EXERCISE 8a

Find each sum.

1. $3\frac{1}{3} + 4\frac{1}{3} =$
2. $10\frac{1}{5} + 1\frac{3}{5} =$
3. $2\frac{5}{12} + 5\frac{5}{12} =$

4. $5\frac{3}{8} + 9\frac{1}{8} =$
5. $8\frac{4}{7} + \frac{2}{7} =$
6. $6\frac{1}{4} + 3\frac{1}{4} =$

7. $1\frac{3}{9} + 2\frac{1}{9} =$
8. $4\frac{1}{11} + 4\frac{9}{11} =$
9. $4\frac{1}{16} + 5\frac{5}{16} + 6\frac{2}{16} =$

10. $6\frac{3}{10} + 4\frac{2}{10} + 5\frac{1}{10} =$
11. $5\frac{1}{3} + 7\frac{1}{3} =$
12. $7\frac{5}{11} + 6\frac{4}{11} + 5\frac{1}{11} =$

13. $9\frac{1}{6} + 3\frac{1}{6} + 2\frac{1}{6} =$
14. $12\frac{3}{8} + 9\frac{2}{8} + 4\frac{1}{8} =$
15. $11\frac{5}{19} + 10\frac{2}{19} + 22\frac{2}{19} =$

WORD PROBLEM

In the following problem, the key word *total* is a clue that you need to add to find the answer.

Shirley built a bookcase and had two pieces of board left. One piece was $11\frac{3}{8}$ inches long, and the other was $4\frac{3}{8}$ inches long. What was the **total** length of board she had left?

Check your answers on page 225.

Simplifying Sums with Fractions Equal to One

Sometimes the fraction in the sum of two or more mixed numbers is improper and equal to 1. Change such a fraction to 1 and add it to the whole number in the sum.

Example 2: Add $6\frac{1}{8} + 3\frac{7}{8}$.

Step 1	Step 2	Step 3	Step 4
$6\frac{1}{8}$	$6\frac{1}{8}$	$6\frac{1}{8}$	$9\frac{8}{8} = 9 + 1 = 10$
$+ 3\frac{7}{8}$	$+ 3\frac{7}{8}$	$+ 3\frac{7}{8}$	
	9	$9\frac{8}{8}$	

Chapter 2: Addition

STEP 1: Set up the problem.

STEP 2: Add the whole numbers: 6 + 3 = 9.

STEP 3: Add the fractions: $\frac{1}{8} + \frac{7}{8} = \frac{8}{8}$.

STEP 4: The sum is $9\frac{8}{8}$. Rewrite the fraction $\frac{8}{8}$ as 1 and add it to the whole number. The sum is 10.

EXERCISE 8b

Find each sum.

1. $3\frac{1}{3} + 4\frac{2}{3} =$
2. $10\frac{1}{6} + 1\frac{5}{6} =$
3. $7\frac{5}{8} + 9\frac{3}{8} =$

4. $6\frac{1}{4} + 4\frac{3}{4} =$
5. $5\frac{5}{7} + 3\frac{2}{7} =$
6. $7\frac{1}{9} + 6\frac{8}{9} =$

7. $12\frac{4}{5} + 6\frac{1}{5} =$
8. $63\frac{1}{2} + \frac{1}{2} =$
9. $11\frac{4}{8} + 9\frac{4}{8} =$

10. $22\frac{11}{16} + 11\frac{5}{16} =$
11. $4\frac{1}{7} + 5\frac{2}{7} + 6\frac{4}{7} =$
12. $2\frac{7}{12} + 3\frac{3}{12} + 8\frac{2}{12} =$

13. $3\frac{3}{16} + 3\frac{6}{16} + 8\frac{7}{16} =$
14. $8\frac{2}{9} + 4\frac{3}{9} + 6\frac{4}{9} =$
15. $32\frac{1}{3} + 13\frac{1}{3} + 21\frac{1}{3} =$

WORD PROBLEM

In the following problem, the key words *in all* are a clue that you need to add to find the answer.

Susan spent $3\frac{1}{2}$ hours watching television on Monday. On Tuesday she watched for $4\frac{1}{2}$ hours. **In all,** how many hours did Susan spend watching television on Monday and Tuesday?

Check your answers on page 225.

Simplifying Sums with Fractions Greater than One

Sometimes the fraction in the sum of two or more mixed numbers is improper and stands for an amount greater than 1. Rewrite such a fraction as a mixed number and add it to the whole number in the sum.

Example 3: Add $3\frac{4}{7} + 2\frac{5}{7}$.

Step 1	Step 2	Step 3	Step 4	Step 5
$3\frac{4}{7}$ $+\ 2\frac{5}{7}$	$3\frac{4}{7}$ $+\ 2\frac{5}{7}$ $\overline{5}$	$3\frac{4}{7}$ $+\ 2\frac{5}{7}$ $\overline{5\frac{9}{7}}$	$7\overline{)9}\ \ 1\frac{2}{7}$ $\ \ \ \ \underline{7}$ $\ \ \ \ 2$	5 $+\ 1\frac{2}{7}$ $\overline{6\frac{2}{7}}$

STEP 1: Set up the problem.
STEP 2: Add the whole numbers: $3 + 2 = 5$.
STEP 3: Add the fractions: $\frac{4}{7} + \frac{5}{7} = \frac{9}{7}$.
STEP 4: Rewrite the improper fraction in the sum as a mixed number: $\frac{9}{7} = 1\frac{2}{7}$.
STEP 5: Add the mixed number to the whole number in the sum: $5 + 1\frac{2}{7} = 6\frac{2}{7}$.

EXERCISE 8c

Find the sum.

1. $6\frac{3}{5} + 4\frac{4}{5} =$
2. $30\frac{3}{4} + 10\frac{3}{4} =$
3. $12\frac{11}{12} + 13\frac{5}{12} =$
4. $8\frac{4}{5} + 3\frac{3}{5} =$
5. $14\frac{7}{9} + 9\frac{5}{9} =$
6. $3\frac{8}{11} + 3\frac{4}{11} =$
7. $9\frac{5}{8} + 4\frac{7}{8} =$
8. $16\frac{4}{5} + 21\frac{4}{5} =$
9. $13\frac{6}{7} + \frac{5}{7} =$
10. $4\frac{5}{6} + \frac{5}{6} =$
11. $4\frac{4}{9} + 6\frac{1}{9} + 8\frac{7}{9} =$
12. $3\frac{9}{16} + 3\frac{1}{16} + 8\frac{11}{16} =$
13. $5\frac{1}{4} + 1\frac{3}{4} + 5\frac{3}{4} =$
14. $7\frac{10}{13} + 5\frac{10}{13} + 3\frac{1}{13} =$
15. $14\frac{19}{25} + 3\frac{22}{25} + 31\frac{10}{25} =$

> **WORD PROBLEM**
>
> In the following problem, look for the key words that tell you to add to find the answer.
>
> On Friday, Manny worked $2\frac{1}{4}$ hours at his second job. On Saturday, he worked $5\frac{3}{4}$ hours more. On Sunday, he put in another $3\frac{3}{4}$ hours. In all how many hours did Manny work at his second job?

Check your answers on page 225.

Lesson 9
Adding Fractions with Unlike Denominators

Fractions can be added only when their denominators are the same. Sometimes, however, you need to add fractions that have different denominators. In such cases, you need to rewrite the fractions using their LCD, as you learned to do in Lesson 4. Then you can add the fractions.

Example: Add $\frac{3}{8} + \frac{7}{12}$.

Step 1

$8 \times 1 = 8 \quad 12 \times 1 = 12$
$8 \times 2 = 16 \quad 12 \times 2 = \mathbf{24}$
$8 \times 3 = \mathbf{24} \quad 12 \times 3 = 36$
$8 \times 4 = 32 \quad$ Etc.
Etc.

Step 2

$\frac{3 \times 3}{8 \times 3} = \frac{9}{24}$

$\frac{7 \times 2}{12 \times 2} = \frac{14}{24}$

Step 3

$\frac{9}{24} + \frac{14}{24} = \frac{23}{24}$

STEP 1: In this case, to find the LCD for the fractions, list some of the multiples of their denominators. The LCD for 8 and 12 is 24.

STEP 2: Raise each fraction to an equivalent fraction with 24 as the denominator.

STEP 3: Using the new fractions, set up the addition problem and add. The sum of $\frac{3}{8}$ and $\frac{7}{12}$ is $\frac{23}{24}$.

Whenever you add fractions with different denominators, try to use the easiest method for finding the LCD. (You learned three ways to find LCDs in Lesson 4.)

EXERCISE 9

Find each sum.

1. $\dfrac{1}{2} + \dfrac{1}{8} =$
2. $\dfrac{3}{5} + \dfrac{3}{10} =$
3. $\dfrac{1}{12} + \dfrac{5}{24} =$

4. $\dfrac{4}{11} + \dfrac{5}{22} =$
5. $\dfrac{5}{9} + \dfrac{1}{3} =$
6. $\dfrac{1}{2} + \dfrac{1}{6} =$

7. $\dfrac{7}{8} + \dfrac{1}{4} =$
8. $\dfrac{3}{4} + \dfrac{1}{2} =$
9. $\dfrac{8}{12} + \dfrac{3}{4} =$

10. $\dfrac{7}{10} + \dfrac{1}{2} =$
11. $\dfrac{1}{4} + \dfrac{4}{15} =$
12. $\dfrac{1}{3} + \dfrac{1}{2} =$

13. $\dfrac{3}{4} + \dfrac{1}{3} =$
14. $\dfrac{4}{7} + \dfrac{2}{3} =$
15. $\dfrac{4}{5} + \dfrac{3}{4} =$

16. $\dfrac{1}{2} + \dfrac{4}{5} =$
17. $\dfrac{5}{6} + \dfrac{3}{8} =$
18. $\dfrac{3}{4} + \dfrac{7}{10} =$

19. $\dfrac{1}{8} + \dfrac{1}{10} =$
20. $\dfrac{3}{10} + \dfrac{5}{12} =$
21. $\dfrac{1}{4} + \dfrac{5}{6} =$

22. $\dfrac{5}{6} + \dfrac{4}{15} =$
23. $\dfrac{3}{8} + \dfrac{5}{16} + \dfrac{1}{4} =$
24. $\dfrac{1}{2} + \dfrac{1}{5} + \dfrac{3}{20} =$

25. $\dfrac{1}{6} + \dfrac{1}{9} + \dfrac{5}{36} =$
26. $\dfrac{1}{3} + \dfrac{1}{6} + \dfrac{1}{12} =$
27. $\dfrac{1}{5} + \dfrac{1}{15} + \dfrac{11}{30} =$

28. $\dfrac{1}{2} + \dfrac{1}{4} + \dfrac{3}{8} =$
29. $\dfrac{1}{3} + \dfrac{1}{4} + \dfrac{5}{12} =$
30. $\dfrac{1}{2} + \dfrac{3}{10} + \dfrac{1}{5} =$

31. $\dfrac{5}{32} + \dfrac{7}{8} + \dfrac{3}{16} =$
32. $\dfrac{1}{3} + \dfrac{1}{4} + \dfrac{1}{6} + \dfrac{1}{12} =$
33. $\dfrac{1}{2} + \dfrac{1}{3} + \dfrac{1}{7} =$

34. $\dfrac{1}{3} + \dfrac{1}{4} + \dfrac{1}{7} =$
35. $\dfrac{1}{4} + \dfrac{1}{3} + \dfrac{1}{5} =$
36. $\dfrac{1}{2} + \dfrac{2}{3} + \dfrac{1}{5} =$

37. $\dfrac{1}{2} + \dfrac{1}{3} + \dfrac{1}{9} =$
38. $\dfrac{1}{2} + \dfrac{2}{3} + \dfrac{1}{4} =$
39. $\dfrac{1}{4} + \dfrac{2}{3} + \dfrac{1}{6} =$

> **WORD PROBLEM**
>
> Look for the key words that tell you to add in the following problem.
>
> When Carrie took up jogging to get some exercise, she ran $\frac{1}{4}$ mile the first day. The second day her muscles ached, so she ran only $\frac{1}{5}$ mile. The third day she felt a little better, so she ran $\frac{1}{2}$ mile. How far did Carrie run in all?

Check your answers on page 225.

Lesson 10: Adding Mixed Numbers with Unlike Denominators

To add mixed numbers whose fractions have different denominators, first find the LCD for the fractions. Then raise the terms of one or more of the fractions so that all the fractions have the same denominator, the LCD. Add the fractions and then the whole numbers. Finally, if necessary, simplify the fraction in the sum to lowest terms.

Example: Add $3\frac{1}{4} + 5\frac{5}{6}$.

Step 1

$3\frac{1 \times 4}{4 \times 3} = 3\frac{3}{12}$

$5\frac{5 \times 2}{6 \times 2} = 5\frac{10}{12}$

Step 2

$3\frac{3}{12}$
$+5\frac{10}{12}$
$\overline{8\frac{13}{12}}$

Step 3

$8\frac{13}{12} = 9\frac{1}{12}$

STEP 1: The LCD for the two fractions is 12. Raise each fraction to an equivalent fraction using 12 as the denominator. $3\frac{1}{4}$ becomes $3\frac{3}{12}$, and $5\frac{5}{6}$ becomes $5\frac{10}{12}$.

STEP 2: Set up the addition problem and add.

STEP 3: Simplify the improper fraction in the sum. Change $\frac{13}{12}$ to $1\frac{1}{12}$ and add it to the whole number: $8 + 1\frac{1}{12} = 9\frac{1}{12}$. The sum of $3\frac{1}{4}$ and $5\frac{5}{6}$ is $9\frac{1}{12}$.

Remember: Whenever you add fractions with different denominators, try to ...siest method for finding the LCD. (You learned three ways to find ...esson 4.)

EXERCISE 10

Find each sum.

1. $3\frac{1}{4} + 4\frac{1}{3} =$
2. $6\frac{1}{3} + 4\frac{1}{2} =$
3. $2\frac{1}{2} + 2\frac{1}{4} =$

4. $5\frac{1}{3} + 3\frac{5}{6} =$
5. $2\frac{1}{2} + 1\frac{3}{4} =$
6. $4\frac{3}{5} + 2\frac{1}{4} =$

7. $5\frac{3}{5} + 2\frac{1}{2} =$
8. $3\frac{3}{8} + 1\frac{2}{5} =$
9. $3\frac{1}{4} + 2\frac{2}{3} =$

10. $6\frac{3}{8} + 2\frac{1}{4} =$
11. $1\frac{7}{8} + 4\frac{9}{16} =$
12. $2\frac{7}{10} + 4\frac{3}{5} =$

13. $4\frac{3}{4} + 4\frac{1}{3} =$
14. $9\frac{1}{3} + 9\frac{4}{7} =$
15. $2\frac{1}{4} + 8\frac{4}{5} =$

16. $6\frac{1}{4} + 8\frac{1}{6} =$
17. $1\frac{2}{3} + 2\frac{1}{4} + 3\frac{1}{6} =$
18. $6\frac{1}{2} + 3\frac{7}{10} + 4\frac{3}{5} =$

19. $4\frac{5}{9} + 6\frac{1}{3} + 10\frac{1}{2} =$
20. $10\frac{1}{4} + 11\frac{1}{2} + 12\frac{5}{6} =$

WORD PROBLEM

Look for the key word that tells you to add to solve the following problem.

Carrie is now training for a road race. One day last week, she ran $2\frac{3}{5}$ miles in the morning and another $4\frac{1}{3}$ miles in the evening. Altogether, how many miles did Carrie run that day?

Check your answers on page 226.

Chapter 3 SUBTRACTION

The three lessons in this chapter cover the subtraction of fractions and mixed numbers. The first two lessons cover the skills needed to subtract without borrowing. The third lesson covers the skills needed when you must borrow before you subtract.

Lesson 11 Subtracting with Like Denominators

In Lessons 7 and 8, you learned to add fractions and mixed numbers with the same denominators. The process for subtraction is similar.

Subtracting Fractions with Like Denominators

To subtract fractions that have the same denominator, subtract the numerator in the smaller fraction from the numerator in the larger fraction. Then write the difference over the denominator. Simplify the answer, if possible.

Example 1: Subtract $\frac{8}{15} - \frac{3}{15}$.

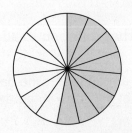 −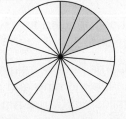

$\frac{8}{15}$ − $\frac{3}{15}$

	Step 1	**Step 2**	**Step 3**
Horizontal Method	$\frac{8}{15} - \frac{3}{15} =$	$\frac{8}{15} - \frac{3}{15} = \frac{5}{15}$	$\frac{5 \div 5}{15 \div 5} = \frac{1}{3}$
Vertical Method	$\begin{array}{r} \frac{8}{15} \\ -\frac{3}{15} \\ \hline \end{array}$	$\begin{array}{r} \frac{8}{15} \\ -\frac{3}{15} \\ \hline \frac{5}{15} \end{array}$	$\frac{5 \div 5}{15 \div 5} = \frac{1}{3}$

44   UNIT 1: Fractions

STEP 1: Set up the problem. (Use either the horizontal or the vertical method.)

STEP 2: Subtract the numerators: 8 − 3 = 5. Write 5 over 15, the denominator. The difference is $\frac{5}{15}$.

STEP 3: Divide 5 into both the numerator and denominator to simplify the answer. The answer to $\frac{8}{15} - \frac{3}{15}$ is $\frac{1}{3}$.

EXERCISE 11a

Subtract the fractions.

1. $\frac{4}{5} - \frac{1}{5} = \frac{3}{5}$
2. $\frac{20}{23} - \frac{9}{23} = \frac{11}{23}$
3. $\frac{3}{5} - \frac{1}{5} = \frac{2}{5}$
4. $\frac{17}{24} - \frac{4}{24} = \frac{13}{24}$
5. $\frac{6}{7} - \frac{4}{7} = \frac{2}{7}$
6. $\frac{15}{16} - \frac{8}{16} = \frac{7}{16}$
7. $\frac{3}{5} - \frac{2}{5} = \frac{1}{5}$
8. $\frac{9}{13} - \frac{1}{13} = \frac{8}{13}$
9. $\frac{11}{12} - \frac{6}{12} = \frac{5}{12}$
10. $\frac{19}{20} - \frac{12}{20} = \frac{7}{20}$
11. $\frac{3}{3} - \frac{1}{3} = \frac{2}{3}$
12. $\frac{7}{15} - \frac{5}{15} = \frac{2}{15}$
13. $\frac{7}{9} - \frac{2}{9} = \frac{5}{9}$
14. $\frac{4}{11} - \frac{1}{11} = \frac{3}{11}$
15. $\frac{5}{7} - \frac{3}{7} = \frac{2}{7}$
16. $\frac{4}{10} - \frac{3}{10} = \frac{1}{10}$
17. $\frac{4}{5} - \frac{2}{5} = \frac{2}{5}$
18. $\frac{6}{7} - \frac{1}{7} = \frac{5}{7}$
19. $\frac{19}{64} - \frac{12}{64} = \frac{7}{64}$
20. $\frac{89}{100} - \frac{18}{100} = \frac{71}{100}$
21. $\frac{7}{8} - \frac{1}{8} = \frac{6}{8} = \frac{3}{4}$
22. $\frac{7}{16} - \frac{5}{16} = \frac{2}{16}$
23. $\frac{9}{15} - \frac{6}{15} = \frac{3}{15} = \frac{1}{3}$
24. $\frac{19}{24} - \frac{13}{24} = \frac{8}{24} = \frac{1}{4}$
25. $\frac{11}{12} - \frac{3}{12} = \frac{8}{12} = \frac{2}{3}$
26. $\frac{4}{9} - \frac{1}{9} = \frac{3}{9} = \frac{1}{3}$
27. $\frac{9}{10} - \frac{5}{10} = \frac{4}{10} = \frac{2}{5}$
28. $\frac{5}{8} - \frac{1}{8} = \frac{4}{8} = \frac{1}{2}$
29. $\frac{11}{12} - \frac{2}{12} = \frac{9}{12} = \frac{3}{4}$
30. $\frac{17}{18} - \frac{7}{18} = \frac{10}{18} = \frac{5}{9}$
31. $\frac{7}{12} - \frac{5}{12} = \frac{2}{12} = \frac{1}{6}$
32. $\frac{13}{15} - \frac{8}{15} = \frac{5}{15} = \frac{1}{3}$
33. $\frac{11}{12} - \frac{1}{12} = \frac{10}{12} = \frac{5}{6}$

34. $\dfrac{3}{8} - \dfrac{1}{8} =$ 35. $\dfrac{20}{36} - \dfrac{18}{36} =$ 36. $\dfrac{17}{30} - \dfrac{9}{30} =$

37. $\dfrac{18}{32} - \dfrac{2}{32} =$ 38. $\dfrac{19}{64} - \dfrac{3}{64} =$ 39. $\dfrac{8}{9} - \dfrac{2}{9} =$

> **WORD PROBLEM**
>
> In the following problem, the key word *left* is a clue that you need to subtract to find the answer.
>
> John had $\dfrac{7}{8}$ yard of copper pipe. He used $\dfrac{3}{8}$ yard to repair a leak. How long was the pipe he had **left**?

Check your answers on page 226.

Subtracting Mixed Numbers with Like Denominators

To subtract mixed numbers whose fractions have the same denominator, first subtract the whole numbers and then subtract the fractions. Simplify the answer, if possible.

Example 2: Subtract $6\dfrac{7}{9} - 4\dfrac{1}{9}$.

Step 1	Step 2	Step 3	Step 4
$6\dfrac{7}{9}$	$6\dfrac{7}{9}$	$6\dfrac{7}{9}$	$2\dfrac{6 \div 3}{9 \div 3} = 2\dfrac{2}{3}$
$-\ 4\dfrac{1}{9}$	$-\ 4\dfrac{1}{9}$	$-\ 4\dfrac{1}{9}$	
	2	$2\dfrac{6}{9}$	

STEP 1: Set up the problem. (Most people prefer to use the vertical method with mixed numbers.)

STEP 2: Subtract the whole numbers: $6 - 4 = 2$.

STEP 3: Subtract the fractions: $\dfrac{7}{9} - \dfrac{1}{9} = \dfrac{6}{9}$.

STEP 4: The difference is $2\dfrac{6}{9}$. Divide the numerator and denominator of the fraction by 3 to simplify it. The answer to $6\dfrac{7}{9} - 4\dfrac{1}{9}$ is $2\dfrac{2}{3}$.

EXERCISE 11b

Subtract the mixed numbers.

1. $16\frac{8}{9} - 8\frac{2}{9} =$
2. $37\frac{4}{5} - 25\frac{2}{5} =$
3. $21\frac{6}{7} - 18\frac{4}{7} =$
4. $65\frac{10}{11} - 45\frac{7}{11} =$
5. $19\frac{2}{3} - 18\frac{1}{3} =$
6. $7\frac{9}{10} - 1\frac{3}{10} =$
7. $6\frac{13}{16} - 2\frac{9}{16} =$
8. $8\frac{11}{36} - 2\frac{7}{36} =$
9. $93\frac{5}{8} - 8\frac{3}{8} =$
10. $203\frac{3}{4} - 74\frac{1}{4} =$
11. $11\frac{3}{4} - 6\frac{1}{4} =$
12. $14\frac{4}{9} - 7\frac{1}{9} =$
13. $25\frac{4}{5} - 1\frac{1}{5} =$
14. $17\frac{5}{6} - 4\frac{1}{6} =$
15. $11\frac{7}{32} - 8\frac{5}{32} =$
16. $32\frac{2}{3} - 17\frac{1}{3} =$
17. $13\frac{4}{5} - 8\frac{1}{5} =$
18. $16\frac{5}{12} - 9\frac{1}{12} =$
19. $88\frac{7}{20} - 79\frac{3}{20} =$
20. $14\frac{7}{10} - 8\frac{2}{10} =$

WORD PROBLEM

Jimmy boxed and wrapped a present to mail to his sister. The package weighed $12\frac{15}{16}$ pounds. The present inside the package weighed $9\frac{3}{16}$ pounds. How much did the packing materials weigh?

Check your answers on page 226.

Lesson 12

Subtracting with Unlike Denominators

In Lessons 9 and 10, you learned that you can add fractions and mixed numbers only when they have the same denominator. Fractions to be subtracted must also have the same denominator.

Subtracting Fractions with Unlike Denominators

To subtract fractions that have different denominators, first find their LCD. (Use one of the three methods for finding LCDs that you learned in Lesson 4.) Then raise one or both fractions to higher terms.

Example 1: Subtract $\frac{1}{4} - \frac{1}{6}$.

Step 1

$4 \times 1 = 4 \quad 6 \times 1 = 6$
$4 \times 2 = 8 \quad 6 \times 2 = \mathbf{12}$
$4 \times 3 = \mathbf{12} \quad 6 \times 3 = 18$
$4 \times 4 = 16 \quad$ Etc.
Etc.

Step 2

$\frac{1 \times 3}{4 \times 3} = \frac{3}{12}$

$\frac{1 \times 2}{6 \times 2} = \frac{2}{12}$

Step 3

$\frac{3}{12} - \frac{2}{12} = \frac{1}{12}$

STEP 1: Find the LCD for the fractions. The lowest multiple common to 4 and 6 is 12.

STEP 2: Raise each fraction to an equivalent fraction with 12 as the denominator.

STEP 3: Set up the problem and subtract. The answer to $\frac{1}{4} - \frac{1}{6} = \frac{1}{12}$.

Be sure to simplify answers whenever you subtract.

EXERCISE 12a

Subtract the fractions.

1. $\frac{3}{4} - \frac{1}{3} =$
2. $\frac{2}{3} - \frac{3}{5} =$
3. $\frac{2}{3} - \frac{1}{2} =$

4. $\frac{1}{6} - \frac{1}{9} =$
5. $\frac{5}{8} - \frac{1}{3} =$
6. $\frac{1}{2} - \frac{1}{5} =$

7. $\frac{7}{9} - \frac{1}{6} =$
8. $\frac{4}{5} - \frac{1}{4} =$
9. $\frac{4}{5} - \frac{1}{2} =$

10. $\frac{5}{8} - \frac{1}{9} =$
11. $\frac{4}{5} - \frac{3}{10} =$
12. $\frac{5}{6} - \frac{2}{3} =$

13. $\frac{9}{10} - \frac{3}{4} =$
14. $\frac{5}{6} - \frac{3}{10} =$
15. $\frac{5}{8} - \frac{1}{6} =$

16. $\dfrac{17}{20} - \dfrac{2}{5} =$

17. $\dfrac{7}{12} - \dfrac{2}{5} =$

18. $\dfrac{7}{8} - \dfrac{1}{3} =$

19. $\dfrac{15}{16} - \dfrac{3}{4} =$

20. $\dfrac{2}{3} - \dfrac{1}{6} =$

WORD PROBLEM

The key words *decreased . . . by* in the following problem are a clue that you need to subtract to find the answer.

For exercise after an operation Carlos walked $\dfrac{3}{5}$ mile each day. After hurting his ankle, he **decreased** his daily distance **by** $\dfrac{1}{10}$ mile. How far did Carlos walk each day after he hurt his ankle?

Check your answers on page 226.

Subtracting Mixed Numbers with Unlike Denominators

To subtract mixed numbers whose fractions have different denominators, first find the LCD for the fractions. Then raise one or both fractions to higher terms. Subtract the whole numbers first, and then subtract the fractions. Finally, simplify the fraction in the answer if necessary.

Example 2: Subtract $4\dfrac{7}{8} - 2\dfrac{1}{6}$.

Step 1	Step 2	Step 3
$4\dfrac{7 \times 3}{8 \times 3} = 4\dfrac{21}{24}$	$4\dfrac{21}{24}$	$4\dfrac{21}{24}$
$2\dfrac{1 \times 4}{6 \times 4} = 2\dfrac{4}{24}$	$-\,2\dfrac{4}{24}$	$-\,2\dfrac{4}{24}$
	$\overline{2}$	$\overline{2\dfrac{17}{24}}$

STEP 1: Find the LCD for the fractions. The lowest multiple common to 8 and 6 is 24. Raise each fraction to an equivalent fraction with 24 as the denominator. $4\dfrac{7}{8}$ becomes $4\dfrac{21}{24}$, and $2\dfrac{1}{6}$ becomes $2\dfrac{4}{24}$.

STEP 2: Set up the problem and subtract the whole numbers.

STEP 3: Subtract the fractions. The answer to $4\dfrac{7}{8} - 2\dfrac{1}{6}$ is $2\dfrac{17}{24}$.

Remember: Be sure to simplify the fraction in answers whenever you subtract.

EXERCISE 12b

Subtract the mixed numbers.

1. $6\frac{5}{8} - 3\frac{1}{4} =$
2. $7\frac{4}{5} - 3\frac{2}{3} =$
3. $9\frac{3}{7} - 2\frac{1}{3} =$

4. $16\frac{4}{5} - 9\frac{3}{4} =$
5. $8\frac{1}{2} - 7\frac{1}{16} =$
6. $12\frac{3}{5} - 7\frac{1}{10} =$

7. $16\frac{5}{6} - 11\frac{1}{4} =$
8. $33\frac{2}{3} - 7\frac{1}{9} =$
9. $17\frac{5}{6} - 4\frac{1}{2} =$

10. $25\frac{1}{2} - 6\frac{1}{8} =$
11. $19\frac{3}{4} - 12\frac{1}{3} =$
12. $132\frac{1}{2} - 76\frac{1}{3} =$

13. $31\frac{9}{16} - 14\frac{1}{4} =$
14. $13\frac{1}{4} - 12\frac{7}{32} =$
15. $111\frac{21}{28} - 38\frac{5}{14} =$

16. $18\frac{3}{5} - 9\frac{1}{3} =$
17. $14\frac{7}{12} - 7\frac{3}{8} =$
18. $116\frac{7}{33} - 47\frac{1}{11} =$

19. $12\frac{3}{4} - 11\frac{2}{5} =$
20. $300\frac{3}{7} - 199\frac{1}{3} =$

WORD PROBLEM

Find the key word in the following problem that tells you to subtract to find the answer.

Mason bought $2\frac{3}{4}$ tons of feed grain. His cows ate $1\frac{1}{6}$ tons the first week. How much grain did Mason have left after the first week?

Check your answers on page 226.

Lesson 13

Borrowing

Sometimes you need to borrow from a whole number before you can subtract.

Subtracting Fractions from the Whole Number 1

To subtract a fraction from 1, first rewrite 1 as a fraction with the same denominator as the one in the fraction you are subtracting. Then subtract.

Example 1: Subtract $1 - \frac{1}{3}$.

Step 1	Step 2	Step 3
1	$1 = \frac{3}{3}$	$\frac{3}{3}$
$-\frac{1}{3}$	$-\frac{1}{3} = \frac{1}{3}$	$-\frac{1}{3}$
		$\frac{2}{3}$

STEP 1: Set up the problem. (Most people prefer to use the vertical method to subtract fractions from whole numbers.)

STEP 2: Rewrite 1 as a fraction with 3 as the denominator.

STEP 3: Subtract the fractions. The answer to $1 - \frac{1}{3}$ is $\frac{2}{3}$.

EXERCISE 13a

Subtract.

1. $1 - \frac{1}{2} =$
2. $1 - \frac{2}{3} =$
3. $1 - \frac{1}{4} =$
4. $1 - \frac{3}{5} =$

5. $1 - \frac{5}{6} =$
6. $1 - \frac{3}{7} =$
7. $1 - \frac{5}{8} =$
8. $1 - \frac{7}{9} =$

9. $1 - \frac{3}{10} =$
10. $1 - \frac{3}{16} =$
11. $1 - \frac{3}{13} =$
12. $1 - \frac{13}{21} =$

> **WORD PROBLEM**
>
> Look for the key word that tells you to subtract to solve the following problem.
>
> Dionne was fixing a door frame. To finish the job, she needed a piece of molding $\frac{13}{16}$ inch long. She had a piece exactly 1 inch long. How much of the 1-inch piece was left after she cut off the part she needed?

Check your answers on page 227.

Subtracting Fractions from Whole Numbers Larger than 1

To subtract a fraction from a whole number larger than 1, borrow 1 from the whole number and rewrite it as a fraction with the same denominator as the one in the fraction you are subtracting.

Example 2: Subtract $3 - \frac{5}{6}$.

Step 1	Step 2	Step 3	Step 4
3	$\overset{2}{\cancel{3}} + 1$	$\overset{2}{\cancel{3}} + 1 = 2\frac{6}{6}$	$2\frac{6}{6}$
$-\frac{5}{6}$	$-\frac{5}{6}$	$-\frac{5}{6} = \frac{5}{6}$	$-\frac{5}{6}$
			$2\frac{1}{6}$

STEP 1: Set up the problem.

STEP 2: Borrow 1 from the whole number 3 and write it in the fraction column of the problem.

STEP 3: Rewrite 1 as a fraction with 6 as the denominator.

STEP 4: Subtract. The answer to $3 - \frac{5}{6}$ is $2\frac{1}{6}$.

EXERCISE 13b

Subtract.

1. $7 - \frac{1}{2} =$
2. $11 - \frac{3}{5} =$
3. $21 - \frac{9}{16} =$
4. $32 - \frac{5}{9} =$

5. $12 - \frac{2}{3} =$
6. $6 - \frac{5}{11} =$
7. $9 - \frac{7}{12} =$
8. $13 - \frac{4}{7} =$

9. $21 - \frac{1}{6} =$ 10. $11 - \frac{1}{8} =$ 11. $3 - \frac{1}{2} =$ 12. $9 - \frac{3}{4} =$

13. $7 - \frac{9}{16} =$ 14. $14 - \frac{5}{6} =$ 15. $27 - \frac{1}{3} =$ 16. $63 - \frac{5}{8} =$

17. $14 - \frac{7}{9} =$ 18. $19 - \frac{3}{5} =$ 19. $41 - \frac{7}{8} =$ 20. $200 - \frac{1}{5} =$

> **WORD PROBLEM**
>
> Find the key word that tells you to subtract in order to solve the following problem.
>
> Tony had 4 gallons of blue paint. He used $\frac{3}{4}$ gallon to paint the shutters on his cottage. How much blue paint did he have left after he painted the shutters?

Check your answers on page 227.

Borrowing When Subtracting Mixed Numbers with Like Denominators

Sometimes you have to borrow in order to subtract the fractions in mixed numbers. This happens when the fraction you are subtracting is the larger fraction.

Example 3: Subtract $7\frac{3}{8} - 4\frac{5}{8}$.

Step 1	Step 2	Step 3	Step 4
$7\frac{3}{8}$	$7\overset{6}{\cancel{8}}\frac{8}{8} + \frac{3}{8}$	$7\overset{6}{\cancel{8}}\frac{8}{8} + \frac{3}{8} = 6\frac{11}{8}$	$6\frac{11}{8}$
$-4\frac{5}{8}$	$-4\ \ \frac{5}{8}$	$-4\ \ \frac{5}{8} = 4\frac{5}{8}$	$-4\frac{5}{8}$
			$2\frac{6}{8} = 2\frac{3}{4}$

STEP 1: Set up the problem. You cannot subtract $\frac{3}{8} - \frac{5}{8}$.

STEP 2: Borrow 1 from the whole number 7 and rewrite it as $\frac{8}{8}$ in the fraction column of the problem.

STEP 3: Combine the two fractions in the top mixed number.

STEP 4: Subtract. Simplify the answer. The answer to $7\frac{3}{8} - 4\frac{5}{8}$ is $2\frac{3}{4}$.

EXERCISE 13c

Subtract.

1. $7\frac{1}{5} - 3\frac{4}{5} =$
2. $4\frac{2}{9} - 2\frac{7}{9} =$
3. $12\frac{1}{6} - 11\frac{5}{6} =$

4. $16\frac{1}{8} - 4\frac{3}{8} =$
5. $5\frac{1}{4} - \frac{3}{4} =$
6. $14\frac{2}{5} - 11\frac{3}{5} =$

7. $9\frac{3}{16} - \frac{15}{16} =$
8. $7\frac{1}{3} - 6\frac{2}{3} =$
9. $2\frac{3}{8} - \frac{5}{8} =$

10. $132\frac{7}{16} - 123\frac{9}{16} =$
11. $5\frac{2}{7} - 4\frac{5}{7} =$
12. $1\frac{1}{4} - \frac{3}{4} =$

> **WORD PROBLEM**
>
> Find the key word that tells you to subtract to solve the following problem.
>
> Alex used $10\frac{7}{8}$ inches of a piece of fabric that was $35\frac{3}{8}$ inches long. How long was the piece of fabric that was left?

Check your answers on page 227.

Borrowing When Subtracting Mixed Numbers with Unlike Denominators

As you learned in Lesson 12, sometimes you have to raise one or both fractions to higher terms before you can subtract. After you do that, it may be necessary to borrow before you can subtract.

Example 4: Subtract $6\frac{1}{3} - 1\frac{2}{5}$.

Step 1

$6\frac{1 \times 5}{3 \times 5} = 6\frac{5}{15}$

$1\frac{2 \times 3}{5 \times 3} = 1\frac{6}{15}$

Step 2

$\overset{5}{\cancel{6}}\frac{15}{15} + \frac{5}{15} = 5\frac{20}{15}$

$- 1 \quad \frac{6}{15} = 1\frac{6}{15}$

Step 3

$5\frac{20}{15}$

$- 1\frac{6}{15}$

$\overline{4\frac{14}{15}}$

54 UNIT 1: Fractions

STEP 1: The lowest multiple that is common for 3 and 5 is 15. Raise each fraction to an equivalent fraction with 15 as the denominator. $6\frac{1}{3}$ becomes $6\frac{5}{15}$, and $1\frac{2}{5}$ becomes $1\frac{6}{15}$. You cannot subtract $\frac{5}{15} - \frac{6}{15}$.

STEP 2: Borrow 1 from the whole number 6 and write it as $\frac{15}{15}$ in the fraction column of the problem. Then combine the two fractions in the top mixed number.

STEP 3: Subtract. The answer to $6\frac{1}{3} - 1\frac{2}{5}$ is $4\frac{14}{15}$.

Remember: Be sure to simplify the fraction in answers whenever you subtract.

EXERCISE 13d

Subtract.

1. $7\frac{2}{3} - 3\frac{4}{5} =$
2. $4\frac{1}{2} - 2\frac{7}{9} =$
3. $12\frac{1}{6} - 3\frac{1}{2} =$

4. $16\frac{1}{9} - 7\frac{5}{6} =$
5. $5\frac{1}{4} - 1\frac{5}{6} =$
6. $14\frac{1}{3} - 7\frac{3}{5} =$

7. $23\frac{1}{2} - 11\frac{3}{4} =$
8. $18\frac{3}{4} - 12\frac{4}{5} =$
9. $64\frac{1}{2} - 33\frac{2}{3} =$

10. $21\frac{1}{5} - 13\frac{7}{10} =$
11. $9\frac{2}{9} - 2\frac{5}{6} =$
12. $7\frac{1}{6} - 3\frac{5}{12} =$

13. $15\frac{1}{3} - 1\frac{1}{2} =$
14. $93\frac{2}{5} - 46\frac{1}{2} =$
15. $77\frac{1}{8} - 38\frac{1}{2} =$

16. $21\frac{1}{2} - 8\frac{3}{4} =$
17. $46\frac{3}{8} - 7\frac{7}{16} =$
18. $136\frac{1}{2} - 123\frac{3}{5} =$

19. $19\frac{1}{4} - 9\frac{4}{5} =$
20. $62\frac{5}{8} - 14\frac{11}{12} =$

> **WORD PROBLEM**
>
> In the following problem, the key words *how many more* are a clue that you need to subtract to find the answer.
>
> Sam hauled $12\frac{1}{4}$ truckloads of trash on Tuesday. On Wednesday he hauled $7\frac{1}{2}$ truckloads. **How many more** truckloads did he haul on Tuesday than on Wednesday?

Check your answers on page 227.

MIXED PRACTICE 1
ADDITION AND SUBTRACTION OF FRACTIONS
AND MIXED NUMBERS

These problems will give you more practice adding and subtracting fractions and mixed numbers. Read each problem carefully and work it.

1. $\frac{3}{7} + \frac{2}{7} =$
2. $\frac{4}{5} - \frac{1}{5} =$
3. $\frac{1}{9} + \frac{2}{9} + \frac{3}{9} =$

4. $\frac{7}{8} - \frac{1}{8} =$
5. $\frac{7}{12} + \frac{5}{12} =$
6. $\frac{9}{10} - \frac{7}{10} =$

7. $1\frac{1}{9} + 2\frac{1}{9} =$
8. $13\frac{4}{5} - 8\frac{1}{5} =$
9. $16\frac{8}{9} - 8\frac{2}{9} =$

10. $4\frac{5}{8} + 3\frac{1}{8} =$
11. $6\frac{3}{10} + 4\frac{1}{10} + 5\frac{1}{10} =$
12. $14\frac{4}{9} - 7\frac{1}{9} =$

13. $3\frac{1}{3} + 4\frac{2}{3} =$
14. $11\frac{1}{2} + \frac{1}{2} =$
15. $8\frac{2}{9} + 4\frac{2}{9} + 6\frac{5}{9} =$

16. $6\frac{3}{5} + 4\frac{4}{5} =$
17. $9\frac{5}{8} + 4\frac{7}{8} =$
18. $5\frac{1}{4} + 1\frac{3}{4} + 5\frac{3}{4} =$

19. $\frac{5}{9} + \frac{1}{3} =$
20. $\frac{2}{3} - \frac{3}{5} =$
21. $\frac{1}{5} + \frac{1}{15} + \frac{11}{30} =$

22. $\frac{1}{3} + \frac{1}{4} + \frac{1}{6} + \frac{1}{12} =$
23. $\frac{1}{3} + \frac{1}{2} =$
24. $\frac{1}{4} + \frac{1}{3} + \frac{1}{5} =$

25. $\frac{1}{2} + \frac{2}{3} + \frac{1}{5} =$
26. $\frac{5}{6} - \frac{3}{10} =$
27. $\frac{1}{10} + \frac{1}{12} =$

56 UNIT 1: Fractions

28. $\dfrac{4}{5} - \dfrac{3}{10} =$

29. $\dfrac{1}{2} + \dfrac{1}{3} + \dfrac{1}{9} =$

30. $\dfrac{5}{8} + \dfrac{3}{10} + \dfrac{3}{4} =$

31. $2\dfrac{1}{2} + 1\dfrac{3}{4} =$

32. $3\dfrac{1}{4} + 2\dfrac{2}{3} =$

33. $6\dfrac{5}{8} - 3\dfrac{1}{4} =$

34. $16\dfrac{5}{6} - 11\dfrac{1}{4} =$

35. $6\dfrac{1}{2} + 3\dfrac{7}{10} + 4\dfrac{3}{5} =$

36. $17\dfrac{5}{6} - 4\dfrac{1}{2} =$

37. $1 - \dfrac{3}{16} =$

38. $3 - \dfrac{4}{7} =$

39. $500 - \dfrac{3}{4} =$

40. $7\dfrac{1}{5} - 3\dfrac{4}{5} =$

41. $12\dfrac{1}{6} - 11\dfrac{5}{6} =$

42. $2\dfrac{3}{8} - \dfrac{5}{8} =$

43. $7\dfrac{2}{3} - 3\dfrac{4}{5} =$

44. $9\dfrac{2}{9} - 2\dfrac{5}{6} =$

45. $21\dfrac{1}{2} - 8\dfrac{3}{4} =$

Check your answers on page 227.

Chapter 4 MULTIPLICATION

The four lessons in this chapter cover the multiplication of fractions and mixed numbers. The first two lessons cover the skills needed to multiply fractions by fractions. The third lesson covers multiplying whole numbers by fractions. The last lesson covers the skills needed to multiply mixed numbers by fractions or by other mixed numbers.

Lesson 14 — Multiplying Fractions by Fractions

When you multiply by a fraction, you find out how much a part of an amount is.

If you want half as many cookies as one recipe makes, you need to use half the amount of each ingredient in the recipe. If the recipe calls for $\frac{3}{4}$ cup of flour, you need to find out what half of that amount is.

$\frac{3}{4}$ cup

$\frac{1}{2}$ of $\frac{3}{4}$ cup

To multiply a fraction by a fraction, multiply the numerators to find the numerator of the product (answer). Then multiply the denominators to find the denominator of the product.

Example: Multiply $\frac{3}{4} \times \frac{1}{2}$.

Step 1	Step 2	Step 3
$\frac{3}{4} \times \frac{1}{2} =$	$\frac{3}{4} \times \frac{1}{2} = \frac{3}{}$	$\frac{3}{4} \times \frac{1}{2} = \frac{3}{8}$

STEP 1: Set up the problem. (Most people prefer the horizontal method for multiplication.)

STEP 2: Multiply the numerators: $3 \times 1 = 3$.

STEP 3: Multiply the denominators: $4 \times 2 = 8$. The product of $\frac{3}{4} \times \frac{1}{2}$ is $\frac{3}{8}$.

EXERCISE 14

Find each product.

1. $\frac{1}{4} \times \frac{1}{4} =$
2. $\frac{2}{3} \times \frac{1}{5} =$
3. $\frac{1}{8} \times \frac{3}{4} =$
4. $\frac{2}{3} \times \frac{1}{7} =$
5. $\frac{1}{3} \times \frac{4}{7} =$
6. $\frac{1}{2} \times \frac{1}{2} =$
7. $\frac{3}{4} \times \frac{3}{20} =$
8. $\frac{1}{2} \times \frac{1}{3} =$
9. $\frac{2}{3} \times \frac{2}{3} =$
10. $\frac{1}{2} \times \frac{1}{5} =$
11. $\frac{3}{4} \times \frac{1}{16} =$
12. $\frac{1}{8} \times \frac{1}{2} =$
13. $\frac{1}{3} \times \frac{1}{9} =$
14. $\frac{2}{3} \times \frac{2}{5} =$
15. $\frac{1}{5} \times \frac{1}{6} =$
16. $\frac{5}{6} \times \frac{1}{3} =$
17. $\frac{1}{6} \times \frac{1}{4} =$
18. $\frac{2}{5} \times \frac{1}{3} =$
19. $\frac{3}{4} \times \frac{3}{4} =$
20. $\frac{1}{3} \times \frac{5}{8} =$

WORD PROBLEM

In the following problem, the key word *of* is a clue that you need to multiply to find the answer.

Carrie had $\frac{2}{3}$ gallon of blue paint. She used $\frac{1}{3}$ **of** it to paint a table and some chairs. What part of a gallon of paint did she use?

Check your answers on page 228.

Lesson 15 — Canceling before Multiplying Fractions

Which of these would be easier to multiply?

$$\frac{15}{16} \times \frac{8}{45} \quad \text{or} \quad \frac{1}{2} \times \frac{1}{3}$$

The second would probably be easier to multiply because it has smaller numbers. Actually, the first was changed into the second by canceling.

You can make many multiplication problems easier by canceling before you multiply.

Canceling Once

To cancel the fractions in a multiplication problem, divide any numerator and any denominator in the problem by the largest number that will divide into both evenly.

Example 1: Multiply $\frac{1}{9} \times \frac{6}{7}$.

Step 1 $\quad \frac{1}{9} \times \frac{6}{7} =$

Step 2 $\quad \frac{1}{\cancel{9}_3} \times \frac{\cancel{6}^2}{7} =$

Step 3 $\quad \frac{1}{\cancel{9}_3} \times \frac{\cancel{6}^2}{7} = \frac{2}{\_}$

Step 4 $\quad \frac{1}{\cancel{9}_3} \times \frac{\cancel{6}^2}{7} = \frac{2}{21}$

STEP 1: Set up the problem.

STEP 2: Cancel by dividing both the denominator 9 and the numerator 6 by 3.

STEP 3: Multiply the resulting numerators: $1 \times 2 = 2$.

STEP 4: Multiply the resulting denominators: $3 \times 7 = 21$. The product of $\frac{1}{9} \times \frac{6}{7}$ is $\frac{2}{21}$.

When you cancel before you multiply fractions, you should not need to simplify the product. However, you should check to see if the product can be simplified.

EXERCISE 15a

Find each product.

1. $\frac{1}{4} \times \frac{4}{8} =$
2. $\frac{9}{10} \times \frac{1}{3} =$
3. $\frac{3}{5} \times \frac{4}{15} =$

4. $\frac{6}{7} \times \frac{2}{3} =$
5. $\frac{1}{6} \times \frac{2}{3} =$
6. $\frac{1}{4} \times \frac{8}{9} =$

7. $\frac{5}{6} \times \frac{3}{4} =$
8. $\frac{8}{11} \times \frac{15}{16} =$
9. $\frac{1}{4} \times \frac{6}{7} =$

10. $\dfrac{5}{8} \times \dfrac{3}{5} =$ 11. $\dfrac{4}{9} \times \dfrac{5}{6} =$ 12. $\dfrac{5}{9} \times \dfrac{3}{4} =$

13. $\dfrac{7}{8} \times \dfrac{2}{3} =$ 14. $\dfrac{8}{9} \times \dfrac{1}{2} =$ 15. $\dfrac{2}{5} \times \dfrac{1}{4} =$

> **WORD PROBLEM**
>
> Look for the key word that tells you to multiply to solve the following problem.
>
> Tony had a job panning gold. His employer allowed him to keep $\dfrac{1}{10}$ of all the gold he found. One day Tony found $\dfrac{15}{16}$ ounce of gold. How much of an ounce was he allowed to keep?

Check your answers on page 228.

Canceling More than Once

In some problems you can cancel more than once before multiplying.

Example 2: Multiply $\dfrac{5}{12} \times \dfrac{8}{20}$.

Step 1	Step 2	Step 3	Step 4	Step 5
$\dfrac{5}{12} \times \dfrac{8}{20} =$	$\dfrac{5}{12} \times \dfrac{\overset{2}{8}}{20} =$	$\dfrac{\overset{1}{5}}{\underset{3}{12}} \times \dfrac{\overset{2}{8}}{\underset{4}{20}} =$	$\dfrac{\overset{1}{5}}{\underset{3}{12}} \times \dfrac{\overset{\overset{1}{2}}{8}}{\underset{\underset{2}{4}}{20}} =$	$\dfrac{\overset{1}{5}}{\underset{3}{12}} \times \dfrac{\overset{\overset{1}{2}}{8}}{\underset{\underset{2}{4}}{20}} = \dfrac{1}{6}$

STEP 1: Set up the problem.

STEP 2: To cancel once, divide both the denominator 12 and the numerator 8 by 4.

STEP 3: To cancel again, divide both the numerator 5 and the denominator 20 by 5.

STEP 4: To cancel a third time, divide both the numerator 2 and the denominator 4 by 2.

STEP 5: Multiply the resulting numerators: $1 \times 1 = 1$. Multiply the resulting denominators: $3 \times 2 = 6$. The product of $\dfrac{5}{12} \times \dfrac{8}{20}$ is $\dfrac{1}{6}$.

When it is possible to cancel more than once, there are often several ways to go about it. For instance, after Step 2 in Example 2, it would be possible to cancel the numerator 2 and the denominator 20. Canceling once again after that, you would end up with the same fractions to multiply: $\dfrac{1}{3} \times \dfrac{1}{2}$.

Chapter 4: Multiplication   61

When you cancel more than once before you multiply fractions, you should not need to simplify the product. However, you should check to see if the product can be simplified.

EXERCISE 15b

Find each product.

1. $\dfrac{9}{10} \times \dfrac{5}{6} =$
2. $\dfrac{8}{25} \times \dfrac{15}{16} =$
3. $\dfrac{2}{3} \times \dfrac{9}{14} =$

4. $\dfrac{2}{5} \times \dfrac{15}{16} =$
5. $\dfrac{4}{9} \times \dfrac{3}{8} =$
6. $\dfrac{7}{8} \times \dfrac{16}{21} =$

7. $\dfrac{7}{12} \times \dfrac{9}{21} =$
8. $\dfrac{4}{20} \times \dfrac{5}{12} =$
9. $\dfrac{3}{7} \times \dfrac{7}{12} =$

10. $\dfrac{3}{5} \times \dfrac{5}{6} =$
11. $\dfrac{2}{9} \times \dfrac{3}{4} =$
12. $\dfrac{28}{35} \times \dfrac{15}{24} =$

WORD PROBLEM

Look for the key word that tells you to multiply to solve the following problem.

Tony's employer liked Tony's dedication to his job, so he took him on as a partner. Now Tony keeps $\frac{4}{9}$ of all the gold he finds. How much of an ounce of gold did Tony keep the day he found $\frac{9}{16}$ ounce?

Check your answers on page 228.

Lesson 16
Multiplying Whole Numbers by Fractions

To multiply a whole number by a fraction, rewrite the whole number as an improper fraction with 1 as the denominator. Then multiply the two fractions. Simplify the product if possible.

Example: Multiply $10 \times \frac{5}{6}$.

Step 1	Step 2	Step 3	Step 4
$10 \times \frac{5}{6} =$	$\frac{10}{1} \times \frac{5}{6} =$	$\frac{\overset{5}{\cancel{10}}}{1} \times \frac{5}{\underset{3}{\cancel{6}}} = \frac{25}{3}$	$\frac{25}{3} = 8\frac{1}{3}$

STEP 1: Set up the problem.

STEP 2: Write the whole number 10 as a fraction over the denominator 1.

STEP 3: Cancel and multiply the fractions.

STEP 4: Simplify the product by changing it to a mixed number, as you learned to do in Lesson 6. The product of $10 \times \frac{5}{6}$ is $8\frac{1}{3}$.

EXERCISE 16

Find the products.

1. $\frac{1}{4} \times 5 =$
2. $\frac{1}{9} \times 3 =$
3. $5 \times \frac{2}{3} =$
4. $4 \times \frac{2}{3} =$
5. $\frac{7}{8} \times 7 =$
6. $\frac{2}{9} \times 4 =$
7. $7 \times \frac{2}{3} =$
8. $\frac{2}{7} \times 6 =$
9. $\frac{3}{8} \times 3 =$
10. $4 \times \frac{3}{5} =$
11. $\frac{4}{5} \times 6 =$
12. $\frac{7}{12} \times 7 =$
13. $27 \times \frac{1}{3} =$
14. $\frac{1}{3} \times 9 =$
15. $4 \times \frac{11}{12} =$

WORD PROBLEM

In the following problem, the key word *rate* is a clue that you need to multiply to solve the problem.

Teresa is an auto mechanic. Her **rate** for labor is $24 per hour. How much would she charge for labor if it took her $\frac{3}{4}$ hour to fix a carburetor?

Check your answers on page 228.

Lesson 17

Multiplying Mixed Numbers by Fractions or Mixed Numbers

Multiplying a mixed number by a fraction or another mixed number is similar to multiplying a whole number by a fraction. First, rewrite each mixed number as an improper fraction. Then multiply the two fractions. Simplify the product if possible.

Example: Multiply $3\frac{2}{3} \times \frac{2}{5}$.

Step 1 $\quad 3\frac{2}{3} \times \frac{2}{5} =$

Step 2 $\quad \frac{11}{3} \times \frac{2}{5} =$

Step 3 $\quad \frac{11}{3} \times \frac{2}{5} = \frac{22}{15}$

Step 4 $\quad \frac{22}{15} = 1\frac{7}{15}$

STEP 1: Set up the problem.

STEP 2: Rewrite the mixed number $3\frac{2}{3}$ as an improper fraction as you learned to do in Lesson 6.

STEP 3: Multiply the fractions.

STEP 4: Simplify the product by rewriting it as a mixed number as you learned to do in Lesson 6. The product of $3\frac{2}{3} \times \frac{2}{5}$ is $1\frac{7}{15}$.

To multiply two mixed numbers, rewrite each of them as an improper fraction. If possible, cancel the fractions before you multiply.

EXERCISE 17

Find the products.

1. $\frac{3}{4} \times 5\frac{1}{2} =$

2. $1\frac{1}{2} \times 2\frac{1}{4} =$

3. $\frac{4}{5} \times 2\frac{1}{3} =$

4. $3\frac{1}{3} \times \frac{4}{9} =$

5. $1\frac{3}{5} \times 4\frac{1}{2} =$

6. $2\frac{1}{2} \times \frac{1}{2} =$

7. $3\frac{1}{2} \times 3\frac{3}{4} =$

8. $1\frac{1}{2} \times 1\frac{1}{2} =$

9. $2\frac{3}{8} \times \frac{4}{19} =$

10. $2\frac{3}{4} \times \frac{8}{33} =$

11. $5\frac{1}{7} \times \frac{7}{9} =$

12. $6\frac{7}{10} \times \frac{1}{2} =$

13. $1\frac{2}{3} \times \frac{2}{3} =$

14. $3\frac{2}{7} \times \frac{1}{4} =$

15. $2\frac{2}{3} \times \frac{1}{4} =$

WORD PROBLEM

In the following problem, the key word *times* is a clue that you need to multiply to solve the problem.

Nancy had a board that was $1\frac{3}{4}$ feet long. For a project she needed another board $1\frac{1}{3}$ **times** as long as the one she had. How long was the board she needed?

Check your answers on page 228.

Chapter 4: Multiplication

Chapter 5
DIVISION

The three lessons in this chapter cover the division of fractions and of combinations of fractions, mixed numbers, and whole numbers.

Lesson 18

Dividing by Fractions

When you divide by a fraction, you find how many parts as large as that fraction there are in another amount.

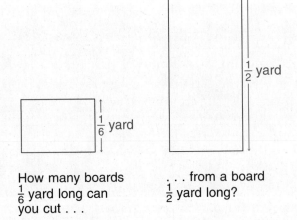

Suppose you need several boards each $\frac{1}{6}$ yard long. You have a board that is $\frac{1}{2}$ yard long. By dividing, you can find out how many shorter pieces you can cut from the longer board.

How many boards $\frac{1}{6}$ yard long can you cut . . .

. . . from a board $\frac{1}{2}$ yard long?

Dividing Fractions by Fractions

To divide a fraction by a fraction, first invert—turn upside down—the fraction you are dividing by. Then multiply the fractions to find the quotient (answer). Simplify the quotient if possible.

Example 1: Divide $\frac{1}{2}$ by $\frac{1}{6}$.

Step 1	Step 2	Step 3	Step 4
$\frac{1}{2} \div \frac{1}{6} =$	$\frac{1}{2} \times \frac{6}{1} =$	$\frac{1}{\cancel{2}} \times \frac{\cancel{6}^{3}}{1} = \frac{3}{1}$	$\frac{3}{1} = 3$

STEP 1: Set up the problem. (Most people prefer the horizontal method for division.)

66 UNIT 1: Fractions

STEP 2: Copy the first fraction, $\frac{1}{2}$. Change the division sign ($\div$) to a multiplication sign ($\times$). Invert the fraction you are dividing by, $\frac{1}{6}$, so that it becomes $\frac{6}{1}$.

STEP 3: Cancel and multiply the fractions.

STEP 4: Simplify the quotient by rewriting the improper fraction $\frac{3}{1}$ as the whole number 3. The answer to $\frac{1}{2} \div \frac{1}{6}$ is 3.

When dividing fractions, cancel only *after* you invert the second fraction. *Never* cancel *before* you invert.

EXERCISE 18a

Find each quotient.

1. $\frac{1}{2} \div \frac{1}{4} =$
2. $\frac{1}{2} \div \frac{1}{8} =$
3. $\frac{1}{6} \div \frac{1}{7} =$

4. $\frac{2}{3} \div \frac{1}{2} =$
5. $\frac{3}{4} \div \frac{2}{3} =$
6. $\frac{4}{7} \div \frac{2}{5} =$

7. $\frac{3}{5} \div \frac{9}{10} =$
8. $\frac{2}{3} \div \frac{4}{9} =$
9. $\frac{7}{8} \div \frac{7}{12} =$

10. $\frac{4}{9} \div \frac{8}{15} =$
11. $\frac{9}{10} \div \frac{2}{5} =$
12. $\frac{1}{2} \div \frac{1}{2} =$

13. $\frac{2}{3} \div \frac{2}{5} =$
14. $\frac{2}{5} \div \frac{2}{3} =$
15. $\frac{5}{8} \div \frac{2}{7} =$

16. $\frac{4}{5} \div \frac{3}{5} =$
17. $\frac{5}{9} \div \frac{5}{9} =$
18. $\frac{3}{8} \div \frac{5}{16} =$

19. $\frac{3}{10} \div \frac{1}{5} =$
20. $\frac{4}{9} \div \frac{2}{3} =$

WORD PROBLEM

In the following problem, the key word *divide* is a clue to the operation you need to use to find the answer.

Denise is landscaping a right-of-way that is $\frac{3}{4}$ mile long. She wants to **divide** the right-of-way into strips each $\frac{1}{8}$ mile long. How many strips can she divide the right-of-way into?

Check your answers on page 228.

Dividing Whole Numbers or Mixed Numbers by Fractions

To divide a whole number or a mixed number by a fraction, rewrite the whole or mixed number as an improper fraction. Then, invert the fraction you are dividing by. Cancel, if you can, and multiply the two fractions. Simplify the quotient if possible.

Example 2: Divide $1\frac{1}{5}$ by $\frac{1}{3}$.

Step 1	Step 2	Step 3	Step 4	Step 5
$1\frac{1}{5} \div \frac{1}{3} =$	$\frac{6}{5} \div \frac{1}{3} =$	$\frac{6}{5} \times \frac{3}{1} =$	$\frac{6}{5} \times \frac{3}{1} = \frac{18}{5}$	$\frac{18}{5} = 3\frac{3}{5}$

STEP 1: Set up the problem.

STEP 2: Rewrite the mixed number $1\frac{1}{5}$ as the improper fraction $\frac{6}{5}$ as you learned to do in Lesson 6.

STEP 3: Copy the first fraction, $\frac{6}{5}$. Change the division sign ($\div$) to a multiplication sign ($\times$). Invert the fraction you are dividing by, $\frac{1}{3}$, so that it becomes $\frac{3}{1}$.

STEP 4: Multiply the fractions.

STEP 5: Simplify the quotient by changing the improper fraction to a mixed number. The answer to $1\frac{1}{5} \div \frac{1}{3}$ is $3\frac{3}{5}$.

Remember: A whole number can be rewritten as an improper fraction with 1 as the denominator.

EXERCISE 18b

Find each quotient.

1. $2 \div \frac{1}{2} =$
2. $4\frac{1}{3} \div \frac{3}{5} =$
3. $8 \div \frac{4}{5} =$

4. $4 \div \frac{2}{3} =$
5. $3\frac{1}{3} \div \frac{6}{7} =$
6. $7\frac{1}{5} \div \frac{7}{10} =$

7. $4 \div \frac{4}{9} =$
8. $3 \div \frac{6}{13} =$
9. $1\frac{1}{2} \div \frac{3}{4} =$

10. $1\frac{3}{10} \div \frac{4}{5} =$
11. $5\frac{1}{4} \div \frac{3}{4} =$
12. $4\frac{3}{8} \div \frac{7}{8} =$

13. $3 \div \frac{2}{3} =$
14. $5\frac{2}{5} \div \frac{3}{10} =$
15. $2 \div \frac{4}{7} =$

16. $6 \div \frac{1}{4} =$

17. $1\frac{3}{10} \div \frac{1}{5} =$

18. $9 \div \frac{3}{4} =$

19. $1\frac{1}{4} \div \frac{5}{6} =$

20. $9 \div \frac{2}{3} =$

> **WORD PROBLEM**
>
> In the following problem, find the key word that tells you to divide to find the answer.
>
> Prasong has a board that is $2\frac{1}{2}$ yards long. How many shelves will he have if he divides the board into pieces that are $\frac{1}{4}$ yard long?

Check your answers on page 229.

Lesson 19: Dividing Fractions or Mixed Numbers by Whole Numbers

To divide a fraction or a mixed number by a whole number, you need to rewrite the whole number as an improper fraction.

Dividing Fractions by Whole Numbers

To divide a fraction by a whole number, first rewrite the whole number as an improper fraction with 1 as its denominator. Then invert the new fraction and multiply.

Example 1: Divide $\frac{1}{2}$ by 3.

Step 1	Step 2	Step 3	Step 4
$\frac{1}{2} \div 3 =$	$\frac{1}{2} \div \frac{3}{1} =$	$\frac{1}{2} \times \frac{1}{3} =$	$\frac{1}{2} \times \frac{1}{3} = \frac{1}{6}$

STEP 1: Set up the problem.

STEP 2: Write the whole number 3 as a fraction with a denominator of 1.

STEP 3: Copy the first fraction, $\frac{1}{2}$. Change the division sign ($\div$) to a multiplication sign ($\times$). Invert $\frac{3}{1}$, the fraction you are dividing by, so that it becomes $\frac{1}{3}$.

STEP 4: Multiply the fractions.

Chapter 5: Division 69

After you invert, see if you can cancel. Remember to simplify quotients if possible.

EXERCISE 19a

Find each quotient.

1. $\frac{2}{3} \div 3 =$
2. $\frac{4}{5} \div 3 =$
3. $\frac{1}{6} \div 2 =$

4. $\frac{7}{8} \div 2 =$
5. $\frac{1}{2} \div 2 =$
6. $\frac{9}{10} \div 3 =$

7. $\frac{1}{5} \div 2 =$
8. $\frac{1}{3} \div 5 =$
9. $\frac{1}{2} \div 7 =$

10. $\frac{2}{5} \div 6 =$
11. $\frac{3}{7} \div 4 =$
12. $\frac{1}{5} \div 5 =$

13. $\frac{2}{3} \div 4 =$
14. $\frac{3}{4} \div 8 =$
15. $\frac{7}{16} \div 4 =$

16. $\frac{2}{3} \div 6 =$
17. $\frac{1}{2} \div 9 =$
18. $\frac{15}{16} \div 7 =$

19. $\frac{2}{3} \div 9 =$
20. $\frac{4}{5} \div 8 =$

WORD PROBLEM

In the following problem, the key word *share* is a clue that you need to divide to find the answer.

Joanne wants to **share** $\frac{3}{4}$ pound of chocolate among 6 people. How much chocolate will each person get?

Check your answers on page 229.

Dividing Mixed Numbers by Whole Numbers

To divide a mixed number by a whole number, first rewrite both numbers as improper fractions. Then invert the fraction you are dividing by and multiply.

Example 2: Divide $1\frac{1}{5}$ by 5.

Step 1	Step 2	Step 3	Step 4
$1\frac{1}{5} \div 5 =$	$\frac{6}{5} \div \frac{5}{1} =$	$\frac{6}{5} \times \frac{1}{5} =$	$\frac{6}{5} \times \frac{1}{5} = \frac{6}{25}$

STEP 1: Set up the problem.

STEP 2: Rewrite the mixed number $1\frac{1}{5}$ as the improper fraction $\frac{6}{5}$. Rewrite the whole number 5 as the improper fraction $\frac{5}{1}$. (Note: You cannot cancel here because you have not yet inverted the second fraction.)

STEP 3: Copy the first fraction, $\frac{6}{5}$. Change the division sign ($\div$) to a multiplication sign ($\times$). Invert the $\frac{5}{1}$, the fraction you are dividing by, so that it becomes $\frac{1}{5}$.

STEP 4: Multiply the fractions.

After you invert, see if you can cancel. Remember to simplify quotients if possible.

EXERCISE 19b

Find each quotient.

1. $2\frac{1}{2} \div 4$
2. $1\frac{3}{8} \div 22$
3. $3\frac{2}{3} \div 11$

4. $4\frac{3}{5} \div 23$
5. $5\frac{1}{4} \div 7$
6. $1\frac{5}{8} \div 26$

7. $2\frac{1}{3} \div 14$
8. $3\frac{2}{5} \div 34$
9. $4\frac{11}{16} \div 25$

10. $2\frac{5}{32} \div 23$
11. $7\frac{2}{7} \div 17$
12. $4\frac{7}{8} \div 13$

WORD PROBLEM

Look for the key word that tells you what operation to use to solve this problem.

Jamela had $1\frac{1}{4}$ pies left after a party. She wanted to share the remaining pie evenly among her 5 children. What part of a pie would each child get?

Check your answers on page 229.

Lesson 20

Dividing by Mixed Numbers

To divide a fraction or a mixed number by a mixed number, you need to rewrite each mixed number as an improper fraction.

Example: Divide $2\frac{2}{3}$ by $1\frac{1}{2}$.

Step 1	Step 2	Step 3	Step 4	Step 5
$2\frac{2}{3} \div 1\frac{1}{2} =$	$\frac{8}{3} \div \frac{3}{2} =$	$\frac{8}{3} \times \frac{2}{3} =$	$\frac{8}{3} \times \frac{2}{3} = \frac{16}{9}$	$\frac{16}{9} = 1\frac{7}{9}$

STEP 1: Set up the problem.

STEP 2: Rewrite both mixed numbers as improper fractions. (Note: You cannot cancel here because you have not yet inverted the second fraction.)

STEP 3: Copy the first fraction, $\frac{8}{3}$. Change the division sign ($\div$) to a multiplication sign ($\times$). Invert the fraction you are dividing by, $\frac{3}{2}$, so that it becomes $\frac{2}{3}$.

STEP 4: Multiply the fractions.

STEP 5: Simplify the quotient by rewriting the improper fraction $\frac{16}{9}$ as the mixed number $1\frac{7}{9}$.

After you invert, see if you can cancel.

EXERCISE 20

Find each quotient.

1. $1\frac{3}{4} \div 1\frac{1}{4} =$
2. $3\frac{1}{2} \div 2\frac{3}{4} =$
3. $1\frac{1}{2} \div 1\frac{1}{2} =$

4. $\frac{3}{8} \div 1\frac{2}{3} =$
5. $3\frac{1}{8} \div 2\frac{1}{2} =$
6. $\frac{2}{3} \div 1\frac{1}{2} =$

7. $\frac{2}{5} \div 1\frac{1}{7} =$
8. $3\frac{3}{4} \div 3\frac{3}{4} =$
9. $2\frac{1}{4} \div 1\frac{1}{4} =$

10. $8\frac{1}{3} \div 1\frac{2}{3} =$
11. $1\frac{3}{4} \div 2\frac{1}{3} =$
12. $\frac{2}{5} \div 4\frac{1}{10} =$

UNIT 1: Fractions

13. $6\dfrac{2}{3} \div 10\dfrac{2}{3} =$

14. $1\dfrac{1}{2} \div 7\dfrac{1}{8} =$

15. $3\dfrac{1}{4} \div 3\dfrac{1}{8} =$

16. $3\dfrac{7}{8} \div 2\dfrac{1}{2} =$

17. $4\dfrac{1}{6} \div 1\dfrac{1}{6} =$

18. $3\dfrac{2}{3} \div 1\dfrac{1}{2} =$

19. $2\dfrac{4}{5} \div 2\dfrac{1}{3} =$

20. $11\dfrac{1}{4} \div 1\dfrac{2}{3} =$

> **WORD PROBLEM**
>
> In the following problem, the key word *each* is a clue that you need to divide to find the answer.
>
> Luz has $13\dfrac{1}{3}$ yards of material to make dresses for bridesmaids. **Each** dress takes $3\dfrac{1}{3}$ yards of material. How many dresses can she make?

Check your answers on page 229.

MIXED PRACTICE 2
MULTIPLICATION AND DIVISION OF FRACTIONS AND MIXED NUMBERS

These problems will give you more practice at multiplying and dividing fractions and mixed numbers. Read each problem carefully and work it.

1. $\dfrac{3}{4} \times \dfrac{1}{2} =$
2. $\dfrac{3}{5} \div \dfrac{1}{2} =$
3. $\dfrac{2}{3} \times \dfrac{1}{5} =$

4. $\dfrac{1}{2} \div \dfrac{1}{4} =$
5. $\dfrac{1}{3} \times \dfrac{4}{7} =$
6. $\dfrac{4}{7} \div \dfrac{2}{5} =$

7. $\dfrac{9}{10} \div \dfrac{2}{5} =$
8. $\dfrac{2}{3} \times \dfrac{2}{3} =$
9. $\dfrac{1}{5} \times \dfrac{1}{6} =$

10. $\dfrac{4}{5} \div \dfrac{3}{5} =$
11. $1\dfrac{1}{5} \div \dfrac{1}{3} =$
12. $\dfrac{1}{9} \times \dfrac{6}{7} =$

13. $2 \div \dfrac{1}{2} =$
14. $\dfrac{9}{10} \times \dfrac{1}{3} =$
15. $\dfrac{3}{5} \times \dfrac{4}{15} =$

16. $7\dfrac{1}{5} \div \dfrac{7}{10} =$
17. $5\dfrac{1}{4} \div \dfrac{3}{4} =$
18. $\dfrac{1}{4} \times \dfrac{6}{7} =$

19. $\dfrac{4}{9} \times \dfrac{5}{6} =$
20. $6 \div \dfrac{1}{4} =$
21. $\dfrac{1}{2} \div 3 =$

22. $\dfrac{5}{12} \times \dfrac{8}{20} =$
23. $\dfrac{2}{3} \div 3 =$
24. $\dfrac{9}{10} \times \dfrac{5}{6} =$

25. $\dfrac{9}{10} \div 3 =$
26. $\dfrac{3}{7} \div 4 =$
27. $\dfrac{4}{5} \times \dfrac{15}{16} =$

28. $\dfrac{2}{3} \div 6 =$
29. $\dfrac{7}{12} \times \dfrac{9}{21} =$
30. $\dfrac{28}{35} \times \dfrac{15}{24} =$

31. $8 \times \dfrac{4}{5} =$
32. $1\dfrac{1}{5} \div 5 =$
33. $2\dfrac{1}{2} \div 4 =$

34. $4\dfrac{3}{5} \div 23 =$
35. $\dfrac{1}{4} \times 5 =$
36. $4 \times \dfrac{2}{3} =$

37. $7 \times \dfrac{2}{3} =$
38. $4\dfrac{11}{16} \div 25 =$
39. $7\dfrac{2}{7} \div 17 =$

40. $\dfrac{7}{12} \times 7 =$
41. $1\dfrac{3}{4} \div 1\dfrac{1}{4} =$
42. $3\dfrac{2}{3} \times \dfrac{2}{5} =$

UNIT 1: Fractions

43. $2\frac{2}{3} \div 1\frac{1}{2} =$
44. $\frac{3}{4} \times 5\frac{1}{2} =$
45. $\frac{2}{3} \div 1\frac{1}{2} =$

46. $3\frac{1}{3} \times \frac{4}{9} =$
47. $3\frac{1}{4} \div 3\frac{1}{8} =$
48. $2\frac{1}{2} \times 3\frac{3}{4} =$

49. $1\frac{2}{3} \times \frac{2}{3} =$
50. $11\frac{1}{4} \div 1\frac{2}{3} =$

Check your answers on page 229.

FRACTION SKILLS REVIEW

Part A. Simplify each fraction.

1. $\frac{4}{6}$
2. $\frac{15}{21}$
3. $\frac{35}{56}$
4. $\frac{18}{72}$
5. $\frac{14}{21}$

Part B. Raise each fraction to higher terms as indicated.

1. Raise $\frac{5}{6}$ to a fraction with a denominator of 12.
2. Raise $\frac{1}{4}$ to a fraction with a denominator of 20.
3. Raise $\frac{3}{8}$ to a fraction with a denominator of 56.
4. Raise $\frac{1}{9}$ to a fraction with a denominator of 72.
5. Raise $\frac{2}{3}$ to a fraction with a denominator of 24.

Part C. Find the larger fraction in each pair.

1. $\frac{7}{10}$ or $\frac{2}{3}$
2. $\frac{3}{8}$ or $\frac{4}{7}$
3. $\frac{5}{8}$ or $\frac{4}{9}$
4. $\frac{5}{6}$ or $\frac{17}{18}$
5. $\frac{5}{6}$ or $\frac{7}{8}$

Part D. Arrange the fractions in each set in order from largest to smallest.

1. $\frac{2}{5}, \frac{1}{3}, \frac{3}{7}$
2. $\frac{5}{16}, \frac{3}{4}, \frac{5}{8}$
3. $\frac{5}{7}, \frac{2}{3}, \frac{3}{4}$

4. $\frac{2}{7}, \frac{3}{8}, \frac{4}{9}$
5. $\frac{1}{3}, \frac{2}{5}, \frac{3}{8}$

Part E. Rewrite each fraction as a mixed number or a whole number.

1. $\frac{3}{2}$
2. $\frac{10}{7}$
3. $\frac{8}{8}$
4. $\frac{12}{8}$
5. $\frac{9}{3}$

Fraction Skills Review

Part F. Rewrite each mixed number as an improper fraction.

1. $2\frac{1}{2}$ 2. $3\frac{1}{4}$ 3. $2\frac{4}{5}$ 4. $2\frac{7}{11}$ 5. $3\frac{2}{9}$

Part G. Solve each problem.

1. $\frac{3}{8} + \frac{2}{8} =$ 2. $\frac{5}{8} - \frac{3}{8} =$ 3. $\frac{1}{3} \times \frac{2}{5} =$

4. $\frac{5}{12} + \frac{1}{12} =$ 5. $\frac{1}{2} \div \frac{1}{4} =$ 6. $\frac{8}{9} + \frac{4}{9} =$

7. $\frac{1}{6} \times \frac{4}{7} =$ 8. $\frac{4}{5} \div \frac{4}{5} =$ 9. $\frac{11}{25} - \frac{9}{25} =$

10. $11\frac{7}{9} - 3\frac{2}{9} =$ 11. $5\frac{1}{6} + 4\frac{5}{6} =$ 12. $14\frac{15}{23} + 9\frac{3}{23} + 8\frac{7}{23} =$

13. $\frac{2}{3} \div \frac{1}{5} =$ 14. $\frac{3}{9} \times \frac{18}{21} =$ 15. $1 - \frac{2}{5} =$

16. $6 - 2\frac{1}{4} =$ 17. $8 \times \frac{3}{4} =$ 18. $\frac{5}{12} \div \frac{3}{4} =$

19. $\frac{1}{9} + \frac{5}{6} =$ 20. $\frac{5}{6} - \frac{1}{4} =$ 21. $\frac{2}{3} + \frac{5}{24} =$

22. $\frac{7}{12} \times 4\frac{1}{2} =$ 23. $\frac{5}{8} \div \frac{5}{12} =$ 24. $7\frac{3}{5} + 3\frac{7}{15} =$

25. $\frac{5}{6} - \frac{1}{9} =$ 26. $\frac{1}{4} \div 5 =$ 27. $1\frac{11}{32} + 2\frac{5}{8} + 4\frac{1}{16} =$

28. $\frac{3}{5} \div 3 =$ 29. $11\frac{2}{21} - 3\frac{3}{7} =$ 30. $21 \div \frac{3}{4} =$

31. $123\frac{1}{24} - 72\frac{1}{8} =$ 32. $3\frac{3}{5} \times 4\frac{1}{6} =$ 33. $2\frac{3}{5} \div 1\frac{11}{15} =$

Check your answers on page 230.

UNIT 1: Fractions

Chapter 6

USING FRACTIONS

In the six lessons in this chapter you will work with measurements; find perimeters, circumferences, and areas; solve problems that contain both fractions and decimals; and round mixed numbers in order to estimate solutions to problems. You will also solve one-step and multistep word problems.

Lesson 21

Working with Measurements Expressed as Fractions or Mixed Numbers

People often express amounts measured in the English system as fractions or mixed numbers. For example, a plumber might buy $8\frac{1}{4}$ feet of pipe, or a tailor might order $2\frac{1}{2}$ yards of material.

Sometimes you need to convert a measurement expressed as a fraction or a mixed number to another form. For instance, a recipe might call for $\frac{3}{4}$ pound of canned cherries. In a store you would probably find canned cherries labeled in ounces. Then it would be useful to know that 12 ounces is the same as $\frac{3}{4}$ pound.

To be able to convert measurements to another form, you need to recall the values of the units of measure in the English system. Use the following table to review some of those values.

UNITS OF MEASURE: THE ENGLISH SYSTEM

Length
12 inches (in.) = 1 foot (ft)
3 feet = 1 yard (yd)
5280 feet = 1 mile (mi)

Time
60 seconds (sec) = 1 minute (min)
60 minutes = 1 hour (hr)

Weight
16 ounces (oz) = 1 pound (lb)
2000 pounds = 1 ton (t)

Liquid
8 fluid ounces (fl oz) = 1 cup (C)
2 cups = 1 pint (pt)
2 pints = 1 quart (qt)
4 quarts = 1 gallon (gal)

Converting Smaller Units of Measure to Larger Units

To convert an amount expressed as a smaller unit of measure to one expressed as a larger unit, divide the number of smaller units you have by the number of those units in one larger unit.

Example 1: Express 4 feet in yards.

Step 1	Step 2	Step 3	Step 4
1 yd = 3 ft	$4 \div 3 =$	$\frac{4}{1} \div \frac{3}{1} =$	$\frac{4}{1} \times \frac{1}{3} = \frac{4}{3} = 1\frac{1}{3}$

STEP 1: Find the number of feet in one yard. There are 3 ft in 1 yd

STEP 2: Set up the problem to divide 4 (the number of feet) by 3 (the number of feet in one yard).

STEP 3: Rewrite the whole numbers as improper fractions.

STEP 4: Copy $\frac{4}{1}$, change the division sign to a multiplication sign, and invert $\frac{3}{1}$. Multiply and simplify your answer. Four feet is the same as $1\frac{1}{3}$ yards.

EXERCISE 21a

Express each of the following measurements in the larger unit indicated.

1. 20 oz = _____ lb

2. 22 fl oz = _____ C

3. 9 qt = _____ gal

4. 54 oz = _____ lb

5. 3960 ft = _____ mi

6. 150 min = _____ hr

7. 45 sec = _____ min

8. 4 oz = _____ lb

9. 7920 ft = _____ mi

10. 37 pt = _____ qt

WORD PROBLEM

To make a tablecloth, Sylvia needs a piece of material 5 feet long. Since cloth is sold by the yard, Sylvia needs to convert the length of the material she needs to yards. How many yards of material does she need?

Check your answers on page 230.

UNIT 1: Fractions

Converting Larger Units of Measure to Smaller Units

Sometimes you need to convert an amount expressed as a larger unit of measure to one expressed as a smaller unit of measure. To do that, multiply the number of larger units by the number of smaller units in one larger unit.

Example 2: Express $1\frac{1}{4}$ cups in fluid ounces.

Step 1	Step 2	Step 3	Step 4
1 C = 8 fl oz	$1\frac{1}{4} \times 8 =$	$\frac{5}{4} \times \frac{8}{1} =$	$\frac{5}{\underset{1}{\cancel{4}}} \times \frac{\overset{2}{\cancel{8}}}{1} = \frac{10}{1} = 10$

STEP 1: Find the number of fluid ounces in one cup. There are 8 fl oz in 1 C.

STEP 2: Set up the problem to multiply $1\frac{1}{4}$ (the number of cups) by 8 (the number of fluid ounces in one cup).

STEP 3: Rewrite the numbers as improper fractions.

STEP 4: Multiply and simplify your answer. $1\frac{1}{4}$ cups is the same as 10 fluid ounces.

EXERCISE 21b

Express each of the following measurements in the smaller unit indicated.

1. $1\frac{1}{2}$ yd = _____ ft

2. $1\frac{1}{2}$ hr = _____ min

3. $4\frac{3}{4}$ lb = _____ oz

4. $8\frac{1}{4}$ gal = _____ qt

5. $5\frac{1}{3}$ ft = _____ in.

6. $4\frac{1}{2}$ gal = _____ qt

7. $3\frac{1}{2}$ mi = _____ ft

8. $12\frac{3}{4}$ gal = _____ qt

9. $2\frac{1}{4}$ C = _____ fl oz

10. $2\frac{1}{8}$ mi = _____ ft

> **WORD PROBLEM**
>
> A lemonade recipe calls for $2\frac{1}{2}$ pints of water. Jamal needs to measure the water in cups. How many cups of water does he need?

Check your answers on page 230.

Converting Units When Two Conversions Are Involved

Sometimes it is necessary to convert between two units of measure whose relationship is not stated in the table on page 79. For example, you may need to convert from inches to yards. The table doesn't show the relationship between inches and yards. However, using the table you can convert from inches to feet and then from feet to yards.

Example 3: Express 60 inches in yards.

Step 1	**Step 2**	**Step 3**
1 ft = 12 in.	$\frac{60}{1} \div \frac{12}{1} =$	$\frac{60}{1} \times \frac{1}{12} = 5$

Step 4	**Step 5**	**Step 6**
1 yd = 3 ft	$\frac{5}{1} \div \frac{3}{1} =$	$\frac{5}{1} \times \frac{1}{3} = \frac{5}{3} = 1\frac{2}{3}$

STEP 1: Find the number of inches in one foot. There are 12 in. in 1 ft.

STEP 2: Set up the problem to divide 60 (the number of inches) by 12 (the number of inches in one foot).

STEP 3: Work the problem. The solution shows that 60 in. equals 5 ft.

STEP 4: Find the number of feet in one yard. There are 3 ft in 1 yd.

STEP 5: Set up the problem to divide 5 (the number of feet in 60 inches) by 3 (the number of feet in one yard).

STEP 6: Work the problem. The solution shows that 5 ft equals $1\frac{2}{3}$ yd. Therefore, 60 in. is the same as $1\frac{2}{3}$ yd.

Solving the problem in Example 3 requires *dividing* twice to convert from a smaller unit of measure to a larger unit and then to an even larger unit. Using two steps to convert a larger unit of measure to a much smaller one requires *multiplying* twice.

EXERCISE 21c

Express each of the following measurements in the unit indicated.

1. 9 gal = _____ pt
2. 60 fl oz = _____ pt
3. $\frac{1}{5}$ hr = _____ sec
4. 84 in. = _____ yd
5. 12 C = _____ gal
6. 360 sec = _____ hr
7. $3\frac{1}{3}$ mi = _____ yd
8. $\frac{1}{2}$ gal = _____ fl oz
9. $\frac{5}{9}$ yd = _____ in.
10. 9000 yd = _____ mi

WORD PROBLEM

Sylvia plans to make another tablecloth, but this time her pattern calls for a piece of cloth 66 inches long. Since she must buy fabric by the yard, how many yards of cloth must she buy?

Check your answers on page 230.

Lesson 22: Finding Perimeters, Circumferences, and Areas

You may recall that the distance around a straight-sided figure (a triangle, a square, or a rectangle) is its perimeter. The distance around a circle is its circumference. The measure of the surface covered by any flat figure is its area. (To review any of these ideas look at Lessons 15 and 32 in *Threshold: Cambridge Pre-GED Program in Mathematics 1.*)

Sometimes a figure's measurements are given in fractions or mixed numbers. To find a perimeter, a circumference, or an area with such measurements, you use the same formulas you use when measurements are in whole numbers or decimals.

Finding Perimeters

To find the perimeter of a figure, add the lengths of all its sides. Remember: To find the perimeter of a square it is faster to multiply since all the sides are the same length. The same is true for triangles with sides of equal length.

Example 1: Find the perimeter of each of the following figures.

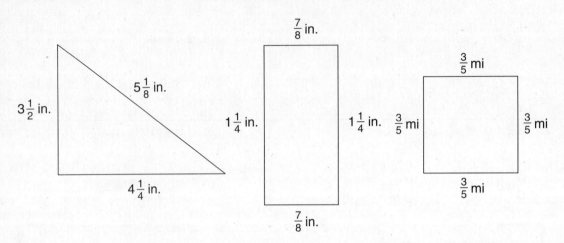

Perimeter: $13\frac{3}{8}$ in.   Perimeter: $4\frac{1}{4}$ in.   Perimeter: $2\frac{2}{5}$ mi

Triangle: To find the perimeter, add

$$3\frac{1}{2} \text{ in.} + 4\frac{1}{4} \text{ in.} + 5\frac{1}{8} \text{ in.} = 13\frac{3}{8} \text{ in.}$$

Rectangle: To find the perimeter, add

$$\frac{7}{8} \text{ in.} + 1\frac{1}{4} \text{ in.} + \frac{7}{8} \text{ in.} + 1\frac{1}{4} \text{ in.} = 4\frac{1}{4} \text{ in.}$$

Square: To find the perimeter, multiply

$$\frac{3}{5} \text{ mi} \times 4 = 2\frac{2}{5} \text{ mi}$$

EXERCISE 22a

Find the perimeter of each of these figures.

1. Rectangle with sides $3\frac{3}{8}$ in. (top and bottom) and $2\frac{1}{2}$ in. (left and right).

2. Triangle with sides $9\frac{5}{6}$ yd, $3\frac{1}{2}$ yd, and $11\frac{1}{4}$ yd.

82   UNIT 1: Fractions

3.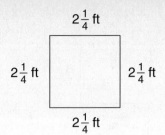

4.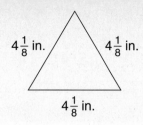

5.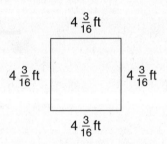

6.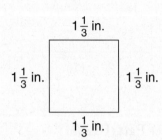

7.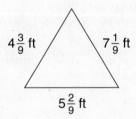

8.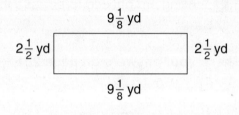

9.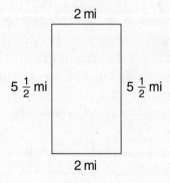

10.

11.

12.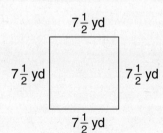

13.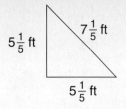

14.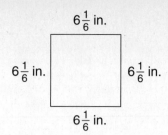

15.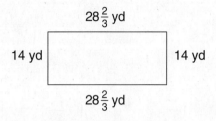

> **WORD PROBLEM**
>
> A rectangular garden measures $3\frac{1}{2}$ yards by $4\frac{1}{4}$ yards. How many yards of fencing are needed to enclose the garden?

Check your answers on page 231.

Finding Circumferences

You may recall that when the measurements of a circle are given in whole numbers or decimals, you use this formula to find its circumference: Circumference = 3.14 × diameter. (To refresh your memory, look at Lesson 32 in *Threshold: Cambridge Pre-GED Program in Mathematics 1*.)

When the measurements of a circle are given in fractions, it is usually easier to use a different form of the same formula:

$$\text{Circumference} = \frac{22}{7} \times \text{diameter}$$

The following example shows how to use this form of the formula.

Example 2: Find the circumference of this circle.

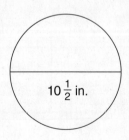

Circle
Circumference: 33 in.

To find the circumference, use this formula:

$$\text{Circumference} = \frac{22}{7} \times \text{diameter}$$

$$\text{Circumference} = \frac{22}{7} \times 10\frac{1}{2} \text{ in.}$$

Circumference = 33 in.

84 UNIT 1: Fractions

EXERCISE 22b

Find the circumference of each of these circles.

1. 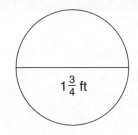 $1\frac{3}{4}$ ft

2. 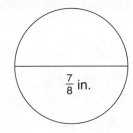 $\frac{7}{8}$ in.

3. 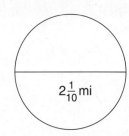 $2\frac{1}{10}$ mi

4. 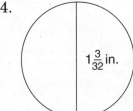 $1\frac{3}{32}$ in.

5. 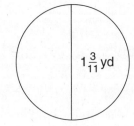 $1\frac{3}{11}$ yd

6. 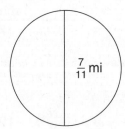 $\frac{7}{11}$ mi

WORD PROBLEM

A child's circular plastic swimming pool has a diameter of $4\frac{3}{8}$ feet. What is the circumference of the pool?

Check your answers on page 231.

Finding Areas

The following example shows the formulas for finding the areas of a rectangle, a square, a triangle, and a circle. Remember that area is expressed in square units of measure.

Example 3: Find the area of each of the following figures.

Rectangle: To find the area, use this formula:

Area = length × width
Area = $\frac{3}{4}$ yd × $1\frac{1}{3}$ yd
Area = 1 sq yd

$1\frac{1}{3}$ yd

$\frac{3}{4}$ yd

Rectangle
Area: 1 sq yd

Chapter 6: Using Fractions

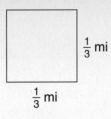

Square
Area: $\frac{1}{9}$ sq mi

Square: To find the area, use this formula:

Area = length × width

Area = $\frac{1}{3}$ mi × $\frac{1}{3}$ mi

Area = $\frac{1}{9}$ mi

Triangle: To find the area, use this formula:

Area = $\dfrac{\text{base} \times \text{height}}{2}$

Area = $\dfrac{\frac{5}{9}\text{mi} \times \frac{3}{10}\text{mi}}{2}$

Area = $\frac{1}{12}$ sq mi

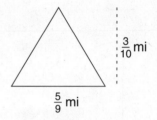

Triangle
Area: $2\frac{2}{5}$ sq mi

Circle
Area: $\frac{11}{14}$ sq ft

Circle: To find the area, use this formula:

Area = $\frac{22}{7}$ × radius × radius

Area = $\frac{22}{7} \times \frac{1}{2}$ ft $\times \frac{1}{2}$ ft

Area = $\frac{11}{14}$ sq ft

EXERCISE 22c

Part A. Use the appropriate formula to find the area of each of the following figures.

1. $3\frac{1}{2}$ in. × $3\frac{1}{2}$ in. (square)

2.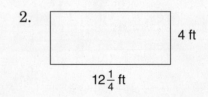
   $12\frac{1}{4}$ ft × 4 ft (rectangle)

86 UNIT 1: Fractions

3.

4.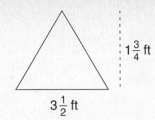

5.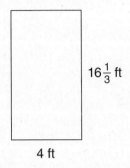

6.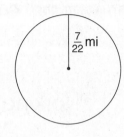

7.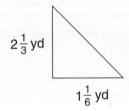

8.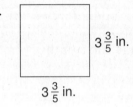

9.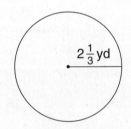

10.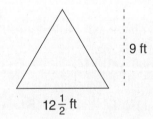

Part B. Use the appropriate formula to find the area of each of the figures described below.

1. A rectangle whose length is 24 inches and whose width is $6\frac{1}{2}$ inches.
2. A square, each of whose sides is $12\frac{2}{3}$ yards long.
3. A triangle whose base is $3\frac{1}{3}$ inches and whose height is $2\frac{5}{6}$ inches.
4. A circle whose radius is $1\frac{9}{33}$ inches long.
5. A square, each of whose sides is $8\frac{3}{4}$ feet long.
6. A rectangle whose length is $12\frac{2}{5}$ miles and whose width is $3\frac{1}{4}$ miles.
7. A circle whose radius is $\frac{1}{10}$ mile long.
8. A triangle whose base is 8 inches and whose height is $3\frac{2}{3}$ inches.
9. A square, each of whose sides is $5\frac{1}{7}$ inches long.
10. A rectangle whose length is $11\frac{3}{8}$ yards and whose width is $6\frac{5}{6}$ yards.

Chapter 6: Using Fractions 87

> **WORD PROBLEM**
>
> A rectangular garden measures $3\frac{1}{2}$ yards by 4 yards. What is the area of the garden?

Check your answers on page 231.

Working with Both Fractions and Decimals

Some problems contain both fractions and decimals. For example, you may want to know the price of $\frac{3}{4}$ pound of potatoes that sell for $.69 per pound.

To solve problems such as these, it is usually easiest to rewrite some of the numbers so that the numbers in the problem are either all fractions or all decimals.

Rewriting Fractions as Decimals

To rewrite a fraction as a decimal, divide its numerator by its denominator.

Example 1: Rewrite $\frac{5}{8}$ as a decimal.

Step 1	Step 2
$\frac{5}{8} = 5 \div 8$	$\begin{array}{r} .625 \\ 8\overline{)5.000} \\ \underline{4\,8} \\ 20 \\ \underline{16} \\ 40 \\ \underline{40} \\ 0 \end{array}$

STEP 1: Set up the division problem so that you divide 5 (the numerator) by 8 (the denominator).

STEP 2: Divide. The result shows that the decimal .625 is equal to the fraction $\frac{5}{8}$.

To refresh your memory about division with decimals, look at Lesson 27 in *Threshold: Cambridge Pre-GED Program in Mathematics 1.*

88   UNIT 1: Fractions

Some fractions equal decimals with many places. For example, $\frac{2}{3}$ equals .666666666 ... (The 6s continue forever.) In most problems if you round the decimal to the thousandths place, your answer will be accurate enough. When the decimal is rounded to the thousandths place, $\frac{2}{3}$ equals .667. (Rounding decimals is covered in Lesson 23 in *Threshold: Mathematics 1*.)

EXERCISE 23a

Rewrite the following fractions as decimals. If necessary, round the decimal to the nearest thousandth.

1. $\frac{1}{5}$
2. $\frac{1}{2}$
3. $\frac{1}{4}$
4. $\frac{3}{4}$
5. $\frac{4}{5}$
6. $\frac{5}{6}$
7. $\frac{3}{10}$
8. $\frac{1}{6}$
9. $\frac{7}{8}$
10. $\frac{2}{9}$
11. $\frac{3}{5}$
12. $\frac{1}{12}$

WORD PROBLEM

Changes in the values of stocks are reported in fractions of a dollar. If you hear that a certain stock went up $\frac{1}{8}$, it means that its value went up $.125 per share. How much does the value of a share of stock decline if it goes down $\frac{3}{8}$?

Check your answers on page 231.

Rewriting Decimals as Fractions

To rewrite a decimal as a fraction, write the value of the decimal as a fraction and simplify.

Example 2: Rewrite .225 as a fraction.

Step 1

$$.225 = \frac{225}{1000}$$

Step 2

$$\frac{225 \div 25}{1000 \div 25} = \frac{9}{40}$$

STEP 1: Write .225 as a fraction.

STEP 2: Divide both parts of the fraction by 25 to simplify it. The decimal .225 equals the fraction $\frac{9}{40}$.

Chapter 6: Using Fractions

EXERCISE 23b

Rewrite the following decimals as fractions.

1. .35
2. .007
3. .018
4. .3
5. .66
6. .75
7. .04
8. .075
9. .2
10. .8
11. .0001
12. .625

WORD PROBLEM

Harry has a drill bit whose diameter is .125 inch. Express the diameter of the bit as a fraction.

Check your answers on page 231.

Solving Problems Containing a Fraction and a Decimal

To solve a problem that contains both a fraction and a decimal, rewrite one of the numbers so that they are both fractions or both decimals.

Example 3: Mildred used $\frac{3}{4}$ of a box of cereal that weighed 1.2 pounds. In pounds, how much cereal did she use?

	Step 1	Step 2
METHOD 1:	$\frac{3}{4} = .75$	1.2 lb × .75 = .9 lb
METHOD 2:	1.2 lb = $1\frac{1}{5}$ lb	$1\frac{1}{5}$ lb × $\frac{3}{4} = \frac{\cancel{6}^{3}}{5} \times \frac{3}{\cancel{4}_{2}} = \frac{9}{10}$ lb

STEP 1: To use Method 1, rewrite the fraction in the problem as a decimal so that both the numbers in the problem are decimals.
To use Method 2, rewrite the decimal in the problem as a fraction so that both the numbers in the problem are fractions.

STEP 2: Using either method, multiply. By Method 1, the answer is .9 lb. By Method 2, the answer is $\frac{9}{10}$ lb. Both answers stand for the same amount. They are equal.

EXERCISE 23c

Multiply to solve each of the following problems.

1. Roger ate $\frac{1}{4}$ of a package of cheese that weighed .8 pound, according to the label. In pounds, how much cheese did he eat? (Hint: This problem can be solved by using either decimals or fractions.)

2. Karin bought $2\frac{3}{4}$ pounds of potatoes at $.80 per pound. How much did she pay? (Hint: Since the answer should be in dollars, it is best to work the problem using decimals.)

3. Mariama can jog 1 mile in $8\frac{1}{2}$ minutes. At the same rate, how long will it take her to jog 3.5 miles? (Hint: This problem can be solved by using either decimals or fractions.)

4. Luz bought $1\frac{3}{4}$ yards of material that cost $5.40 per yard. How much did Luz pay for the cloth? (Hint: Since the answer should be in dollars, it is best to work the problem using decimals.)

5. Beng Choo wanted only $\frac{1}{4}$ of the 3.6 tons of feed that she bought to be delivered the next day. How many tons of feed did she receive the next day? (Hint: This problem can be solved by using either decimals or fractions.)

Check your answers on page 231.

Rounding Mixed Numbers and Estimating Solutions

As you will see in Lesson 25, the next-to-last step in solving a word problem is estimating its solution. When the problem involves mixed numbers, it is usually helpful to round each mixed number to the nearest whole number before estimating the solution.

Rounding Mixed Numbers

To round a mixed number, compare the fraction in that number to $\frac{1}{2}$. If the fraction is smaller than $\frac{1}{2}$, drop it. If the fraction is equal to or larger than $\frac{1}{2}$, increase the whole number by 1 and drop the fraction.

Example 1: Round these two mixed numbers: $4\frac{3}{8}$ and $9\frac{5}{8}$.

Step 1	Step 2	Step 3
$\dfrac{1 \times 4}{2 \times 4} = \dfrac{4}{8}$	The fraction in $4\frac{3}{8}$ is smaller than $\frac{4}{8}$.	Rounded, $4\frac{3}{8}$ is 4.
	The fraction in $9\frac{5}{8}$ is larger than $\frac{4}{8}$.	Rounded, $9\frac{5}{8}$ is 10.

STEP 1: The fractions in both mixed numbers have 8 as their denominator. So that you can compare the fractions to $\frac{1}{2}$, rewrite $\frac{1}{2}$ as a fraction with 8 as the denominator.

STEP 2: Compare the fraction in each mixed number to $\frac{4}{8}$.

STEP 3: Because the fraction in $4\frac{3}{8}$ is smaller than $\frac{4}{8}$, round the mixed number by dropping the fraction. Because the fraction in $9\frac{5}{8}$ is larger than $\frac{4}{8}$, round the mixed number by increasing the whole number by 1 (9 + 1 = 10) and dropping the fraction.

To compare some fractions to $\frac{1}{2}$, you will need to multiply by 2 to raise the terms of the original fraction. For example, $\frac{4}{9}$ must be raised to $\frac{8}{18}$ and then compared to $\frac{9}{18}$ because $\frac{1}{2}$ cannot be raised to a fraction with 9 as the denominator. This applies to any fraction whose denominator is an odd number.

EXERCISE 24a

Round each mixed number to the nearest whole number.

1. $8\frac{5}{8}$
2. $23\frac{2}{3}$
3. $12\frac{3}{10}$
4. $7\frac{1}{2}$

5. $17\frac{3}{8}$
6. $14\frac{7}{11}$
7. $9\frac{3}{5}$
8. $30\frac{1}{4}$

9. $15\frac{3}{7}$
10. $6\frac{5}{6}$
11. $12\frac{5}{12}$
12. $99\frac{7}{9}$

13. $55\frac{2}{5}$
14. $67\frac{1}{3}$
15. $45\frac{5}{8}$

> **WORD PROBLEM**
>
> Jack had to prepare an estimate in whole hours of the time it would take him to complete a certain job. He remembered that a similar job had once taken him exactly $6\frac{1}{4}$ hours. How much time should Jack estimate for this job?

Check your answers on page 232.

Estimating Solutions

To estimate the solution to a problem that involves mixed numbers, round each mixed number to the nearest whole number and work the problem using the whole numbers—in your head, if possible.

Example 2: Estimate the solution to $15\frac{3}{7} + 3\frac{5}{8}$.

Step 1	**Step 2**
Rounded, $15\frac{3}{7}$ is 15.	$15 + 4 = 19$
Rounded, $3\frac{5}{8}$ is 4.	

STEP 1: Round each mixed number. Because $\frac{3}{7}$ is smaller than $\frac{1}{2}$ (that is, $\frac{6}{14}$ is smaller than $\frac{7}{14}$), round $15\frac{3}{7}$ to 15. Because $\frac{5}{8}$ is larger than $\frac{1}{2}$ (or $\frac{4}{8}$), round $3\frac{5}{8}$ to 4.

STEP 2: Estimate the answer to the problem by adding the rounded mixed numbers. (The actual answer to the problem, without rounding, is $19\frac{3}{56}$, which is very close to 19.

Note: This method of estimating may *not* be useful with problems that involve only proper fractions because any proper fraction would round to either 0 or 1.

EXERCISE 24b

Estimate the solution to each of the following problems.

1. $13\frac{5}{7} + 8\frac{1}{5}$
2. $5\frac{3}{4} \div 2\frac{1}{6}$
3. $2\frac{7}{10} \times 4\frac{7}{11}$

Chapter 6: Using Fractions 93

4. $25\frac{2}{9} - 15\frac{1}{3}$ 5. $9\frac{5}{8} \times 6\frac{3}{10}$ 6. $48\frac{1}{4} \div 7\frac{5}{7}$

7. $30\frac{2}{7} + 70\frac{2}{5}$ 8. $36\frac{2}{3} - 16\frac{5}{7}$ 9. $8\frac{7}{8} \div 2\frac{5}{6}$

10. $82\frac{11}{12} - 40\frac{3}{11}$ 11. $4\frac{1}{8} \times 25\frac{1}{6}$ 12. $19\frac{7}{9} + 20\frac{3}{8}$

13. $38\frac{3}{4} \div 12\frac{7}{10}$ 14. $50\frac{3}{7} \times 19\frac{1}{2}$ 15. $54\frac{1}{9} - 45\frac{5}{11}$

WORD PROBLEM

In the morning William had $18\frac{1}{4}$ bolts of cloth. During the day he sold $7\frac{1}{2}$ bolts. What was his estimate of the number of bolts he had left?

Check your answers on page 232.

Lesson 25: Solving Word Problems with Fractions and Mixed Numbers

The five steps for solving fractions and mixed-number word problems are the same as those for solving word problems with whole numbers and decimals. As a quick reminder, the steps are listed below:

- Understand the question.
- Find the facts you need to solve the problem.
- Decide which math operation to use.
- Estimate the solution.
- Solve the problem and check your answer.

The following example illustrates the steps in a problem involving mixed numbers.

Example: Altogether, Anita needs $2\frac{3}{4}$ bags of concrete mix to patch her 20-foot driveway. She already has $1\frac{1}{8}$ bags. How much more concrete mix does Anita need?

STEP 1: Understand the question.

94 UNIT 1: Fractions

Before you can solve a word problem, you must be sure you know what the problem asks you to find. Usually it helps to do two things: (1) form a mental picture of the problem, and (2) state the problem in different words.

You could picture and restate the example problem this way:

Anita needs to use $2\frac{3}{4}$ bags of concrete mix.   She has $1\frac{1}{8}$ bags.

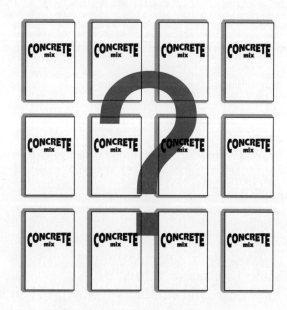

How many more bags of concrete mix does Anita need?

The pictures and the restatement can help you see that the problem asks you to find this: How much more concrete mix than she already has will Anita use?

STEP 2: Find the facts you need to solve this problem.

The next thing to do is to find the facts, or numbers, you must use to solve the problem. Sometimes a problem contains "extra" facts: numbers you don't need to use in solving the problem.

Chapter 6: Using Fractions 95

In the example, you need two facts: (1) the number of bags of concrete mix Anita will use, and (2) the number of bags she already has. You don't need the length of the driveway to solve the problem.

You **need** this fact.
↓

Altogether Anita needs **$2\frac{3}{4}$ bags** of concrete mix to patch her **20-foot** driveway. She already has **$1\frac{1}{8}$ bags**.
↑ ↑
You **don't need** You **need** this fact.
this "extra" fact.

STEP 3: Decide which math operation to use.

The third step is to decide which arithmetic operation to use. Ask yourself, "Do I need to add, subtract, multiply, or divide?" Key words in a problem can help you select the correct operation, as can your mental picture of the problem.

The question part of the example problem has key words that tell you to subtract.

How much more concrete mix does Anita need?
↑
These key words are a clue
that you need to **subtract**.

The pictures of the problem on page 97 suggest that you need to find the *difference* between the amount of concrete mix Anita needs to use and the amount she already has. To find a difference, you need to subtract.

STEP 4: Estimate the solution.

Estimating the answer to a problem can help you decide whether you have chosen the right operation. If the estimated answer makes sense, you have probably chosen correctly. Later, you can compare your final solution to your estimate. The two should be close in value.

To estimate an answer, round the facts in the problem and perform the required math operation. (You practiced rounding mixed numbers and estimating solutions in Lesson 24.)

To estimate the solution to the example problem, think of the problem this way:

3 bags
Altogether, Anita needs $2\frac{3}{4}$ ~~bags~~ of concrete mix to patch her

1 bag
20-foot driveway. She already has $1\frac{1}{8}$ ~~bags~~.

Then subtract:

3 bags − 1 bag = 2 bags

It makes sense that Anita would need about 2 more bags of concrete mix. Therefore, the final solution to the problem should be close to 2 bags.

STEP 5: Solve the problem and check your answer.

To solve the problem, work it using the actual facts from the problem. Read the example problem again.

> Altogether, Anita needs $2\frac{3}{4}$ **bags** of concrete mix to patch her 20-foot driveway. She already has $1\frac{1}{8}$ **bags**. How much more concrete mix does Anita need?

Solve the problem this way:

$$\begin{aligned} 2\tfrac{3}{4} \text{ bags} &= 2\tfrac{6}{8} \text{ bags} \\ -\ 1\tfrac{1}{8} \text{ bags} &= 1\tfrac{1}{8} \text{ bags} \\ \hline &\ 1\tfrac{5}{8} \text{ bags} \end{aligned}$$

Anita needs another $1\frac{5}{8}$ bags of concrete mix to patch her driveway. Notice that the answer would round to 2 bags, the estimated solution.

Finally, check your answer:

$$\begin{aligned} &\ 1\tfrac{5}{8} \text{ bags} \\ +&\ 1\tfrac{1}{8} \text{ bags} \\ \hline &\ 2\tfrac{6}{8} \text{ bags} = 2\tfrac{3}{4} \text{ bags} \end{aligned}$$

Checking shows that the solution is correct.

To refresh your memory more thoroughly about using these steps, look over Lessons 18 and 28 in *Threshold: Cambridge Pre-GED Program in Mathematics 1.*

EXERCISE 25

Solve each problem.

1. John worked $\frac{1}{2}$ a day on Friday and harvested $1\frac{1}{2}$ tons of wheat. On Saturday, he worked almost a full day and harvested another $2\frac{3}{8}$ tons. How much wheat did John harvest altogether?

2. Emily has about 300 corn plants planted in rows. She bought 9 bags of fertilizer for them. Each row of plants needs $2\frac{1}{4}$ bags. How many rows can Emily fertilize?

3. Earl wants to make 4 metal shelves for his workshop. Each shelf is to be $9\frac{1}{4}$ feet long. How much metal shelving does Earl need?

4. Last week Andrea worked $3\frac{3}{4}$ days on a construction job. She also spent $2\frac{1}{2}$ days painting her house. Altogether, how many days did Andrea work that week?

5. Fred needs to make 12 jump ropes for the students in his exercise class. He needs $3\frac{1}{6}$ yards of rope for each jump rope. How many yards of rope does Fred need?

6. Peggy speared 2 fish while SCUBA diving. One was a bluefish that weighed 6 pounds. The other, a blackfish, weighed $\frac{3}{4}$ pound less than the bluefish. How much did the blackfish weigh?

7. Before Joan had a computer, it took her $3\frac{1}{2}$ hours to write her 10-page weekly report. Now, with a computer, it takes her only $1\frac{3}{4}$ hours each week. How much time does the computer save Joan on each weekly report?

8. Blanca bought $3\frac{1}{4}$ yards of screening material from a design store. Her neighbor, Adrienne, bought $12\frac{1}{2}$ yards. Altogether, how many yards of screening material did the 2 neighbors buy?

9. Kathy had $4\frac{3}{4}$ gallons of tan paint. She used $\frac{1}{2}$ of it to paint 3 rooms in her apartment. How much paint did Kathy use?

10. Dimitri cleans houses for a living. He charges $12 per hour. If he spends $3\frac{1}{4}$ hours cleaning a 4-room house, how much does Dimitri earn?

11. Ted made $\frac{1}{2}$ gallon of ice cream. He wants to share it among 4 people. How much ice cream will each person get?

12. Celia divided $2\frac{2}{3}$ quarts of ice cream into portions that were each $\frac{1}{6}$ quart. How many portions did Celia have?

13. Juan works for a furniture manufacturer. It takes him $1\frac{1}{4}$ hours to sand and prime a table top. How many hours does it take Juan to sand and prime 5 table tops?

14. Felix had 3 partly-used cans of paint thinner. One can was $\frac{1}{3}$ full, another was $\frac{1}{4}$ full, and the last was $\frac{1}{6}$ full. What portion of a can of paint thinner did Felix have altogether?

15. To hold about 500 books, Sing put 9 bookshelves on his wall. Each shelf is $3\frac{1}{3}$ feet long. How many feet of bookshelves does Sing have altogether?

16. Joanna's stew recipe calls for $1\frac{3}{4}$ pints of beef stock. She has only $\frac{7}{8}$ pint. How much more beef stock does Joanna need?

17. Miranda used $2\frac{1}{2}$ pounds of ground beef to make 10 hamburgers. How many pounds did each hamburger weigh?

18. Antonio covered 18 stairs with rubber matting. He had bought 45 feet of matting, but he used only $40\frac{1}{2}$ feet. How much matting did Antonio have left over?

19. Ahmed's juicer squeezed 7 oranges and made $2\frac{1}{2}$ cups of juice in 1 minute. How many minutes would it take the juicer to make $13\frac{3}{4}$ cups of orange juice?

20. Martha ordered 15 dozen rolls of paper towels for the shelter where she works. Only $12\frac{5}{12}$ dozen were delivered. How many dozen rolls of paper towels were not delivered?

Check your answers on page 232.

Lesson 26

Solving Multistep Word Problems with Fractions and Mixed Numbers

To solve many word problems, you need to perform more than one operation. For example, you may need to add first and then multiply. Often you need to add, subtract, multiply, or divide to find a fact (a number) you need to solve a problem.

Example: Andre designed two dresses. One requires $2\frac{1}{3}$ yards of material, and the other requires $3\frac{1}{4}$ yards. The material Andre wants for the dresses will cost $12 per yard. How much will enough material for both dresses cost?

Step 1

$$2\frac{1}{3} \text{ yd} = 2\frac{4}{12} \text{ yd}$$
$$+ \ 3\frac{1}{4} \text{ yd} = 3\frac{3}{12} \text{ yd}$$
$$\overline{\phantom{+ \ 3\frac{1}{4} \text{ yd} =} 5\frac{7}{12} \text{ yd}}$$

Step 2

$$5\frac{7}{12} \text{ yd} \times \$12 = \frac{67}{12} \times \frac{12}{1} = \$67$$

STEP 1: To find how much material Andre needs, add $2\frac{1}{3}$ yd + $3\frac{1}{4}$ yd. He needs $5\frac{7}{12}$ yd.

STEP 2: To find out the cost of the material, multiply $5\frac{7}{12}$ yd × $12. The material will cost $67 altogether.

EXERCISE 26

Part A. Read each problem carefully. Choose the answer that tells which operations you would need to do to solve each problem.

1. Louise works at a part-time job for $3\frac{1}{2}$ hours each day. She earns $7 per hour. How much money does she earn in 5 days?
   (1) divide then multiply
   (2) multiply then divide
   (3) multiply then multiply
   (4) add then multiply
   (5) multiply then subtract

Chapter 6: Using Fractions

2. Jonah bought $6\frac{1}{4}$ yards of shelving at $12 a yard. How much change did he receive from $100?

   (1) add then divide
   (2) multiply then subtract
   (3) divide then multiply
   (4) add then multiply
   (5) divide then subtract

3. A field measures $2\frac{1}{4}$ miles by $3\frac{1}{3}$ miles. It takes 300 sacks of clover seed to cover 1 square mile of a field. How many sacks of seed will it take to cover the entire field?

   (1) multiply then divide
   (2) multiply then multiply
   (3) divide then divide
   (4) divide then multiply
   (5) add then multiply

4. Sarah had a piece of electrical wire that was $8\frac{1}{4}$ feet long. She had another piece that was 2 times as long as the first piece. How many feet of wire did she have altogether?

   (1) divide then add
   (2) multiply then divide
   (3) multiply then subtract
   (4) subtract then multiply
   (5) multiply then add

5. The length of one of Daisy's steps when she runs is $\frac{3}{4}$ yard. How many steps does she take when she runs around this park one time?

   (1) subtract then multiply
   (2) multiply then divide
   (3) add then multiply
   (4) add then divide
   (5) multiply then multiply.

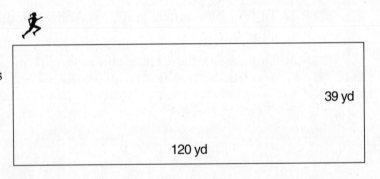

Part B. Choose the answer that gives the solution to each problem.

1. Louise works at a part-time job for $3\frac{1}{2}$ hours each day. She earns $7 per hour. How much money does she earn in 5 days?

   (1) $ 2.50
   (2) $ 17.50
   (3) $ 24.50
   (4) $ 35.00
   (5) $122.50

100   UNIT 1: Fractions

2. Jonah bought $6\frac{1}{4}$ yards of shelving at $12 a yard. How much change did he receive from $100?

   (1) $ .52
   (2) $25.00
   (3) $75.00
   (4) $88.00
   (5) $93.75

3. A field measures $2\frac{1}{4}$ miles by $3\frac{1}{3}$ miles. It takes 300 sacks of clover seed to cover 1 square mile of a field. How many sacks of seed will it take to cover the entire field?

   (1) 2250
   (2) 1675
   (3) $202\frac{1}{2}$
   (4) $7\frac{1}{2}$
   (5) $5\frac{7}{12}$

4. Sarah had a piece of electrical wire that was $8\frac{1}{4}$ feet long. She had another piece that was 2 times as long as the first piece. How many feet of wire did she have altogether?

   (1) $4\frac{1}{8}$ feet
   (2) $12\frac{3}{8}$ feet
   (3) $16\frac{1}{2}$ feet
   (4) $24\frac{3}{4}$ feet
   (5) 33 feet

5. The length of one of Daisy's steps when she runs is $\frac{3}{4}$ yard. How many steps does she take when she runs around this park one time?

   (1) $238\frac{1}{2}$
   (2) 318
   (3) 424
   (4) 3600
   (5) 4800

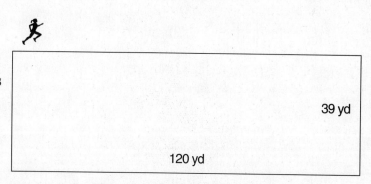

Part C. Solve each of the following problems.

1. The original price of a lamp was $39. During a sale the lamp sold for $\frac{2}{3}$ of the original price. Glen bought 2 of the lamps. How much did he pay altogether?

2. A rectangular pool cover measures $5\frac{1}{3}$ yards by 9 yards. At $4 per square yard, what is the cost of the pool cover?

Chapter 6: Using Fractions   101

3. Millen used $10\frac{1}{2}$ gallons of gasoline to drive 210 miles. She has another 160 miles to drive. How many more gallons of fuel will she use?

4. Everything in Sal's houseware store was on sale for $\frac{3}{4}$ of the original price. Jim bought the articles listed below at the sale price. How much did he pay in all?

Articles Jim Bought	Original Price
toaster	$24
mop	$ 8
set of pots	$96

5. Mino wants to tile the floor of this room. He plans to use tiles that measure $1\frac{1}{3}$ feet on each side. How many tiles does he need in all? (*Hint*: This is a three-step problem.)

16 ft

12 ft

Check your answers on page 233.

FRACTIONS REVIEW

Solve each problem.

1. $\dfrac{3}{20} + \dfrac{7}{20} + \dfrac{11}{20} =$
2. $\dfrac{3}{4} \times \dfrac{1}{2} =$
3. $\dfrac{8}{9} - \dfrac{5}{9} =$

4. $\dfrac{4}{9} \div \dfrac{2}{3} =$
5. $4\dfrac{7}{10} - \dfrac{3}{10} =$
6. $5\dfrac{1}{4} + 1\dfrac{3}{4} + 5\dfrac{3}{4} =$

7. $\dfrac{2}{3} - \dfrac{1}{4} =$
8. $\dfrac{7}{8} \times \dfrac{2}{3} =$
9. $12\dfrac{3}{4} - 2\dfrac{2}{5} =$

10. $\dfrac{1}{2} + \dfrac{1}{3} + \dfrac{1}{9} =$
11. $1 - \dfrac{5}{9} =$
12. $1\dfrac{1}{4} \div \dfrac{5}{6} =$

13. $3 - \dfrac{1}{3} =$
14. $\dfrac{3}{5} \times \dfrac{5}{6} =$
15. $1\dfrac{1}{4} + 1\dfrac{1}{2} + 2\dfrac{1}{6} =$

16. $7\dfrac{3}{8} - 4\dfrac{5}{8} =$
17. $4 \times \dfrac{11}{12} =$
18. $\dfrac{4}{5} \div 8 =$

19. $2\dfrac{2}{3} \times \dfrac{1}{4} =$
20. $4\dfrac{7}{8} \div 13 =$
21. $6\dfrac{1}{3} - 5\dfrac{2}{5} =$

22. $2\dfrac{4}{5} \div 2\dfrac{1}{3} =$
23. $210 \text{ min} = \underline{\quad} \text{ hr}$
24. $1\dfrac{3}{4} \text{ gal} = \underline{\quad} \text{ qt}$

Questions 25 to 32 are based on the following figures.

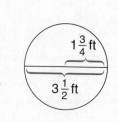

25. What is the perimeter of the triangle?
26. What is the area of the triangle?
27. What is the perimeter of the rectangle?
28. What is the area of the rectangle?

29. What is the perimeter of the square?

30. What is the area of the square?

31. What is the circumference of the circle?

32. What is the area of the circle? _____

33. Hector needs to drill a hole large enough to insert a peg that is .625 inch in diameter. His drill bits are labeled in fractions. What size drill bit should Hector use to make the hole?

34. Minerva bought $6\frac{1}{2}$ yards of material to make dresses for her granddaughters. Each dress requires $1\frac{5}{8}$ yards of material. How many dresses can Minerva make?

35. Ali jogs 1 mile in 8 minutes. How long does it take him to run $\frac{1}{4}$ of a 3.5-mile course?

Check your answers on page 233.

GED PRACTICE 1

This section will give you practice in answering questions like those on the GED. The Mathematics Test of the GED has 56 multiple-choice questions. Each question has 5 choices. The 15 questions in this Practice are all multiple-choice like the ones on the GED. As you do this Practice, use the skills you've studied in this unit—especially the skills for solving word problems with fractions and mixed numbers.

- Understand the question.
- Find the facts you need to solve the problem.
- Decide which math operation to use.
- Estimate the solution.
- Solve the problem and check your answer.

Directions: Choose the one best answer to each item.

1. Frank mixed $\frac{1}{4}$ gallon of blue paint with $1\frac{1}{2}$ gallons of white paint. How many gallons of paint did he have altogether?
   - (1) $\frac{1}{6}$
   - (2) $\frac{3}{8}$
   - (3) $1\frac{1}{4}$
   - (4) $1\frac{3}{4}$
   - (5) 6

2. Bob's cake recipe calls for $\frac{5}{8}$ cup of sugar. He has $\frac{3}{8}$ cup. How many more cups of sugar does Bob need?
   - (1) $\frac{15}{64}$
   - (2) $\frac{1}{4}$
   - (3) $\frac{3}{5}$
   - (4) 1
   - (5) $1\frac{2}{3}$

3. Connie has $6\frac{3}{4}$ pounds of chocolate. If she gives $2\frac{1}{8}$ pounds to Hugh, how many pounds of chocolate will she have left?
   - (1) $3\frac{3}{17}$
   - (2) $4\frac{1}{2}$
   - (3) $4\frac{5}{8}$
   - (4) $8\frac{7}{8}$
   - (5) $14\frac{11}{32}$

4. On Tuesday Fran ran $4\frac{1}{4}$ miles. She decreased her distance on Wednesday by $1\frac{1}{2}$ miles. How many miles did she run on Wednesday?
   - (1) $6\frac{3}{8}$
   - (2) $5\frac{3}{4}$
   - (3) $3\frac{3}{4}$
   - (4) $2\frac{5}{6}$
   - (5) $2\frac{3}{4}$

5. Doron had $\frac{4}{5}$ quart of ginger ale. He used $\frac{1}{4}$ of it in a punch. How many quarts of ginger ale did he use in the punch?
   - (1) $\frac{1}{5}$
   - (2) $\frac{15}{16}$
   - (3) $1\frac{1}{20}$
   - (4) $3\frac{1}{5}$
   - (5) 5

6. David usually spends $\frac{3}{4}$ of his 8-hour workday in meetings. How many hours does David spend each day in meetings?
   - (1) $\frac{3}{32}$
   - (2) 6
   - (3) $7\frac{1}{4}$
   - (4) $8\frac{3}{4}$
   - (5) $10\frac{2}{3}$

7. Lisa worked $5\frac{1}{4}$ hours on Saturday. She spent $\frac{1}{3}$ of the time working on a computer. In hours, how much time did she spend on the computer?
   (1) $\frac{4}{7}$
   (2) $1\frac{3}{4}$
   (3) $4\frac{11}{12}$
   (4) $5\frac{7}{12}$
   (5) $15\frac{3}{4}$

8. Marilyn had $5\frac{1}{3}$ yards of fabric. She divided it into $\frac{2}{3}$-yard pieces to make throw pillows. How many pillows could she make?
   (1) $3\frac{5}{9}$
   (2) $4\frac{2}{3}$
   (3) $5\frac{1}{3}$
   (4) 6
   (5) 8

9. Liam has $\frac{3}{4}$ hour left before quitting time. He still has 9 boxes to wrap. If he divides his time equally, how many hours can he take to wrap each box?
   (1) $\frac{1}{12}$
   (2) $6\frac{3}{4}$
   (3) $8\frac{1}{4}$
   (4) $9\frac{3}{4}$
   (5) 12

10. To get a better price a group of friends ordered 100 pounds of beef. They divided the beef up equally so that each person's share was $12\frac{1}{2}$ pounds. How many people were in the group?
    (1) 8
    (2) 12
    (3) 88
    (4) 113
    (5) 125

11. Hermiña needs $1\frac{1}{2}$ pints of canned chicken broth to make soup. How many 12-ounce cans of broth must she use?
    (1) 1
    (2) $1\frac{1}{2}$
    (3) 2
    (4) 3
    (5) 4

12. Regulations say that there must be at least $1\frac{1}{2}$ square yards of space per person in this pool. How many people are allowed in the pool at one time?
    (1) 22
    (2) 44
    (3) 49
    (4) 231
    (5) 579

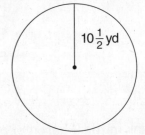

13. Sammy bought a new screwdriver that was on sale for $\frac{3}{4}$ of the original price of $4.80. How much did he pay?
    (1) $ .36
    (2) $1.20
    (3) $3.60
    (4) $4.05
    (5) $6.40

14. Maggie's new family room measures $12\frac{1}{2}$ feet by 16 feet. She can have the floor tiled at a cost of $3 per square foot. What will the tiling cost?
    (1) $ 48
    (2) $ 85.50
    (3) $171
    (4) $200
    (5) $600

15. One small can of paint will cover 46 square feet. How many cans will it take to paint a wall that measures 8 feet by $11\frac{1}{2}$ feet?
    (1) $1\frac{4}{19}$
    (2) 2
    (3) $19\frac{1}{2}$
    (4) 38
    (5) 92

Check your answers on page 233.

GED PRACTICE 1 SKILLS CHART

To review the mathematics skills covered by the items in GED Practice 1, study the following lessons in Unit 1.

Unit 1		Item Number
Lesson 10	Adding Mixed Numbers with Unlike Denominators	1
Lesson 11	Subtracting with Like Denominators	2
Lesson 12	Subtracting with Unlike Denominators	3
Lesson 13	Borrowing	4
Lesson 15	Canceling Before Multiplying Fractions	5
Lesson 16	Multiplying Whole Numbers by Fractions	6
Lesson 17	Multiplying Mixed Numbers by Fractions or Mixed Numbers	7
Lesson 18	Dividing by Fractions	8
Lesson 19	Dividing Fractions and Mixed Numbers by Whole Numbers	9
Lesson 20	Dividing Fractions by Mixed Numbers	10
Lesson 21	Working with Measurements Expressed as Fractions or Mixed Numbers	11
Lesson 22	Finding Perimeters, Circumferences, and Areas	12, 14, 15
Lesson 23	Working with Both Fractions and Decimals	13
Lesson 25	Solving Word Problems with Fractions and Mixed Numbers	(1–10, 13)
Lesson 26	Solving Multistep Word Problems with Fractions and Mixed Numbers	(11–12, 14–15)

UNIT 2

Percents

Chapter 1 of this unit introduces percents and their relationships to decimals and fractions. In Chapter 2 you will learn to use formulas for solving the three basic kinds of percent problems. In Chapter 3 you will learn to estimate answers to percent problems and to solve one-step and multistep percent word problems.

Unit 2 Overview

Chapter 1 Percents, Decimals, and Fractions
Chapter 2 Solving Percent Problems
Chapter 3 Working with Percents

GED Practice 2

Chapter 1

PERCENTS, DECIMALS, AND FRACTIONS

The six lessons in this chapter introduce percents and their equivalent decimals and fractions. You will learn to change percents to decimals and fractions and to change fractions and decimals to percents. These lessons cover some of the skills you will use in the other two chapters in this unit.

Lesson 27

Understanding Percents

Like a fraction or a decimal, a **percent** stands for part of a whole amount. The word *percent* means "per 100." The sign for percent is %.

Writing Percents

A percent refers to a whole that can be divided into 100 equal parts. It describes a portion of that whole.

This square is divided into 100 equal parts.
20 of the parts are shaded, so the shaded portion is 20% of the whole.
80 of the parts are unshaded, so the unshaded portion is 80% of the whole.

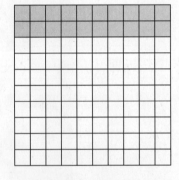

Shaded portion: 20%
Unshaded portion: 80%

Because a dollar can be divided into 100 equal parts, portions of a dollar can be described as percents of a dollar.

A penny is 1% of a dollar.   A nickel is 5% of a dollar.   A nickel and two pennies are 7% of a dollar.

110   UNIT 2: Percents

EXERCISE 27a

Part A. For each figure, tell what percent is shaded and what percent is unshaded.

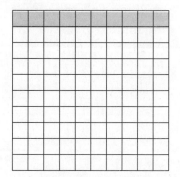

1. (a) shaded portion _____
   (b) unshaded portion _____

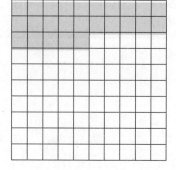

2. (a) shaded portion _____
   (b) unshaded portion _____

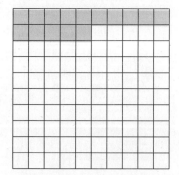

3. (a) shaded portion _____
   (b) unshaded portion _____

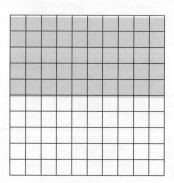

4. (a) shaded portion _____
   (b) unshaded portion _____

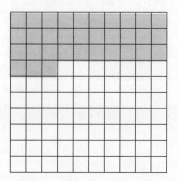

5. (a) shaded portion _____
   (b) unshaded portion _____

6. (a) shaded portion _____
   (b) unshaded portion _____

Part B. Tell what percent of a dollar each of the following drawings represents.

1. _____ % of $1   2. _____ % of $1

3. _____ % of $1

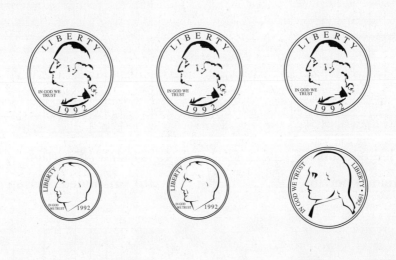

4. _____ % of $1

WORD PROBLEM

Write the percent that answers the following question.

One hundred people registered to take a math class at a community college. On the first night only 88 of the students showed up. What percent of the class was present the first night?

Check your answers on page 234.

Percents Larger than 100%

Percents can be used to represent portions larger than one whole.

These large squares are each divided into 100 equal parts.

120 of the parts are shaded.

The shaded portion is 120% of one large square.

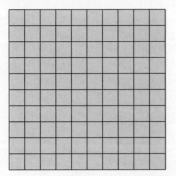 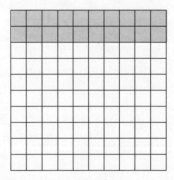

**Shaded portion:
120% of one large square**

EXERCISE 27b

For each figure, tell what percent of one large square is shaded.

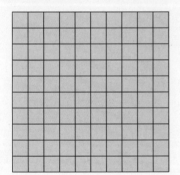 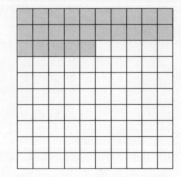

1. shaded portion _____

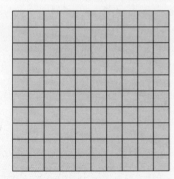 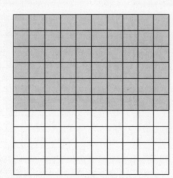

2. shaded portion _____

Chapter 1: Percents, Decimals, and Fractions   113

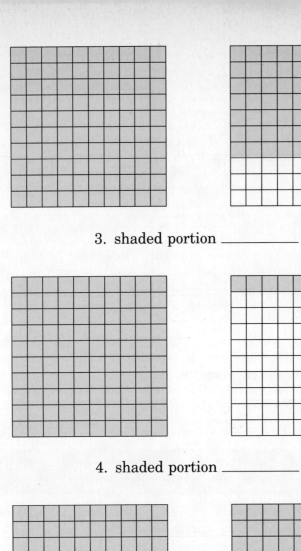

3. shaded portion _____

4. shaded portion _____

5. shaded portion _____

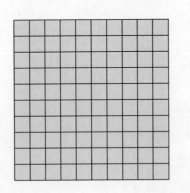 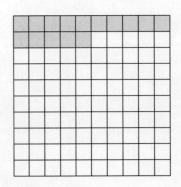

6. shaded portion _____

UNIT 2: Percents

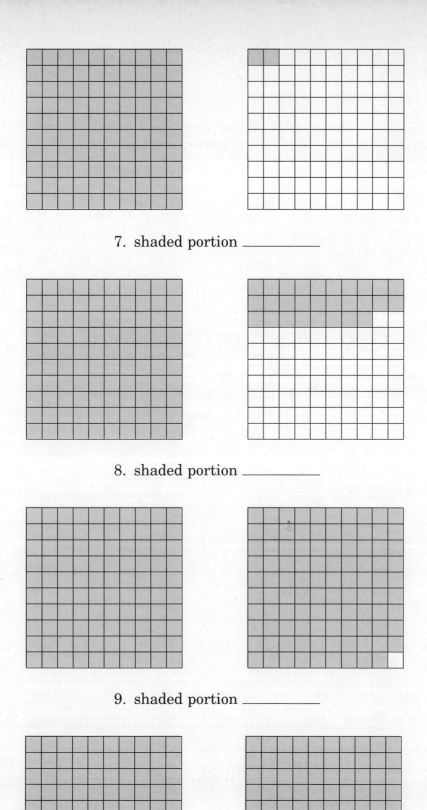

7. shaded portion _____

8. shaded portion _____

9. shaded portion _____

10. shaded portion _____

Chapter 1: Percents, Decimals, and Fractions

> **WORD PROBLEM**
>
> Write the percent that answers the following question.
>
> Jerry deposited $100 in a savings account. Because Jerry's account earned interest, at the end of one year he had $105. What percent of his original $100 deposit did Jerry have at the end of one year?

Check your answers on page 234.

Lesson 28: Rewriting Percents as Decimals

A percent stands for part of a whole amount just as a decimal does.

This large square is divided into 100 equal parts. 35 of the parts are shaded.

The shaded portion can be described as 35% of the whole.

It can also be described as .35 (thirty-five hundredths) of the whole.

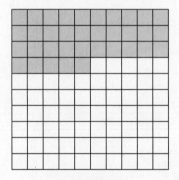

Shaded portion: 35% or .35

As you will see in later lessons, to solve some percent problems you need to rewrite a percent as a decimal.

Moving the Decimal Point

To rewrite a percent as a decimal move the decimal point in the percent two places to the left and drop the percent sign. Drop any zeros at the end of the decimal.

Example 1: Rewrite 50% as a decimal.

Step 1	Step 2	Step 3
50.%	.50.% = .50	.5̶0̶ = .5

116   UNIT 2: Percents

STEP 1: Write the percent. If you like, include the decimal point so that you can see where you are moving it from.

STEP 2: Move the decimal point two places to the left and drop the percent sign.

STEP 3: Drop the zero at the end of the decimal. 50% equals .5 (five tenths).

EXERCISE 28a

Rewrite each percent as a decimal.

1. 55% 2. 32% 3. 67% 4. 73%

5. 44% 6. 20% 7. 10% 8. 88%

9. 70% 10. 60% 11. 14% 12. 90%

13. 12% 14. 30% 15. 76% 16. 99%

17. 40% 18. 59% 19. 80% 20. 41%

WORD PROBLEM

Write the decimal called for by the problem.

Gull Wing Airways had an on-time arrival rate of 87% this month. Express that rate using a decimal.

Check your answers on page 234.

Using Zeros as Place Holders

When you move the decimal point in a percent to rewrite as a decimal, you may have to use a zero to hold a decimal place.

Example 2: Rewrite 4% as a decimal.

　　　　　　　Step 1　　　**Step 2**
　　　　　　　　4.%　　　　.04.% = .04

STEP 1: Write the percent with its decimal point.

STEP 2: Add a zero (0) before the 4. Then move the decimal point two places to the left. Drop the percent sign.

EXERCISE 28b

Rewrite each percent as a decimal.

1. 5%   2. 8%   3. 2%   4. 9%

5. 1%   6. 7%   7. 3%   8. 6%

> **WORD PROBLEM**
>
> Write the decimal that answers the following question.
>
> When Harriet took a written test to get her driver's license, she answered 5% of the questions incorrectly. Expressed as a decimal, how many of her answers were wrong?

Check your answers on page 234.

Rewriting Percents as Mixed Decimals

When you move the decimal point two places to the left in some percents, the result is a mixed decimal.

Example 3: Rewrite 132% as a decimal.

$$\begin{array}{cc} \text{Step 1} & \text{Step 2} \\ 132.\% & 1.32.— = 1.32 \end{array}$$

STEP 1: Write the percent with its decimal point.

STEP 2: Move the decimal point two places to the left and drop the percent sign. The result is a mixed decimal.

EXERCISE 28c

Rewrite each percent as a decimal.

1. 111%   2. 231%   3. 453%   4. 312%

5. 768%   6. 176%   7. 325%   8. 575%

9. 625%   10. 312%   11. 989%   12. 211%

13. 909%   14. 854%   15. 458%   16. 356%

17. 201%   18. 102%   19. 607%   20. 653%

> **WORD PROBLEM**
>
> Write the decimal that answers the following question.
>
> Years ago Stan made a deposit of $100 in a savings fund at his job. His account has grown by 250% since then. Write a mixed decimal to express how many times Stan's account has increased.

Check your answers on page 234.

Rewriting Decimals as Percents

To solve some problems it is necessary to rewrite a decimal as a percent. For example, you may use decimals to work a problem but use a percent to express the answer.

Moving the Decimal Point

To rewrite a decimal as a percent move the decimal point two places to the right and add a percent sign. Drop any zeros to the left of the whole number in the percent. Drop the decimal point in the percent if it is not needed.

Example 1:   Rewrite .06 as a percent.

Step 1	Step 2	Step 3
.06	.06.% = 06.%	06.% = 6%

STEP 1: Write the decimal.

STEP 2: Move the decimal point two places to the right and add a percent sign.

STEP 3: Drop the zero to the left of the whole number. Because the decimal point is not needed, drop it. The decimal .06 (six hundredths) equals 6%.

Chapter 1: Percents, Decimals, and Fractions   119

EXERCISE 29a

Rewrite each decimal as a percent.

1. .45
2. .67
3. .02
4. .09
5. .75
6. .07
7. .44
8. .08
9. .01
10. .33
11. .54
12. .78
13. .98
14. .12
15. .23
16. .66
17. .76
18. .89
19. .03
20. .08

> **WORD PROBLEM**
>
> Write the percent called for by the problem.
>
> José had a delivery of home heating oil in February. In two weeks he used .11 of the oil. What percent of the oil did he burn during those weeks?

Check your answers on page 234.

Rewriting Decimals with More than Two Places as Percents

If a decimal has more than two places, you may need to keep the decimal point when you rewrite the decimal as a percent.

Example 2: Rewrite .375 as a percent.

        Step 1   **Step 2**

        .375   .37.5% = 37.5%

STEP 1: Write the decimal.

STEP 2: Move the decimal point two places to the right and add a percent sign. Do not drop the decimal point because it is a necessary part of the percent. The decimal .375 (three hundred seventy-five thousandths) equals 37.5%.

NOTE: You can read 37.5% as "thirty-seven point five percent" or as "thirty-seven and one half percent."

UNIT 2: Percents

EXERCISE 29b

Rewrite each decimal as a percent.

1. .645
2. .845
3. .075
4. .767

5. .024
6. .333
7. .289
8. .751

9. .889
10. .032
11. .297
12. .789

13. .8765
14. .0043
15. .0504
16. .111

17. .312
18. .7355
19. .1415
20. .0777

WORD PROBLEM

Write the percent called for by the problem.

In the past ten years, an average of .445 of all applicants were accepted for enrollment in a certain vocational school. What percent of the applicants were accepted?

Check your answers on page 234.

Rewriting Decimals with Only One Place as Percents

If a decimal has only one place, you will need to add a zero when you change the decimal to a percent.

Example 3: Rewrite .3 as a percent.

Step 1	Step 2	Step 3
.3	.30.% = 30.%	30.% = 30%

STEP 1: Write the decimal.

STEP 2: Add a zero (0) after the 3 in the decimal and move the decimal point two places to the right. Add a percent sign.

STEP 3: Drop the decimal point because it is not necessary. The decimal .3 (three tenths) equals 30%.

EXERCISE 29c

Rewrite each decimal as a percent.

1. .2
2. .5
3. .7
4. .9
5. .1
6. .4
7. .6
8. .8

> **WORD PROBLEM**
>
> Write the percent called for by the problem.
>
> Althea spends .2 of her earnings for food. What percent of her earnings is spent for food?

Check your answers on page 235.

Rewriting Percents as Fractions

In Lesson 27 you learned that *percent* means "per 100." A percent stands for a portion of something that can be divided into 100 parts. Any percent can be changed to a fraction with a denominator of 100. Sometimes the fraction can be simplified.

This large square is divided into 100 equal parts. 35 of the parts are shaded.

The shaded portion can be described as 35% of the whole.

It can also be described as $\frac{35}{100}$ of the whole, which can be simplified to $\frac{7}{20}$.

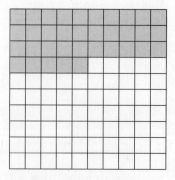

Shaded portion:
As a percent: 35%
As a fraction: $\frac{35}{100}$ or $\frac{7}{20}$

122 UNIT 2: Percents

Rewriting Percents without Decimal Places as Fractions or Mixed Numbers

To rewrite a percent that has no decimal places as a fraction, write the percent as a fraction with a denominator of 100. Drop the percent sign. Simplify the fraction, if possible.

Example 1: Rewrite 40% as a fraction.

Step 1	Step 2	Step 3
40%	$40\% = \dfrac{40}{100}$	$\dfrac{40}{100} = \dfrac{2}{5}$

STEP 1: Write the percent.

STEP 2: Write the percent as a fraction with a denominator of 100. Drop the percent sign.

STEP 3: To simplify the fraction, divide both 40 and 100 by 20. $40\% = \dfrac{2}{5}$.

When a percent is larger than 100%, it can be changed to a mixed number.

Example 2: Rewrite 240% as a mixed number.

Step 1	Step 2	Step 3
240%	$240\% = \dfrac{240}{100}$	$\dfrac{240}{100} = 2\dfrac{40}{100} = 2\dfrac{2}{5}$

STEP 1: Write the percent.

STEP 2: Write the percent as a fraction with a denominator of 100. Drop the percent sign.

STEP 3: Simplify the fraction.

EXERCISE 30a

Rewrite each percent as a fraction simplified to lowest terms.

1. 30%
2. 80%
3. 510%
4. 90%
5. 70%
6. 325%
7. 75%
8. 50%
9. 60%
10. 55%
11. 112%
12. 95%

Chapter 1: Percents, Decimals, and Fractions

13. 208%   14. 4%   15. 96%   16. 35%

17. 45%   18. 40%   19. 320%   20. 5%

> **WORD PROBLEM**
>
> This word problem gives you a percent but asks you to express the answer in a fraction.
>
> This year The Sanderson School expects to enroll more than 1000 adult students. The faculty expects 25% of the students to be new to the school. What fraction of the students will be new to the school?

Check your answers on page 235.

Rewriting Percents with Decimal Places as Fractions or Mixed Numbers

Sometimes you need to rewrite a percent with decimal places as a fraction. First rewrite the percent as a decimal. Then rewrite the decimal as a fraction and simplify.

Example 3: Rewrite 62.5% as a fraction.

$$\text{Step 1} \qquad \text{Step 2} \qquad \text{Step 3}$$
$$62.5\% = .625 \qquad .625 = \frac{625}{1000} \qquad \frac{625}{1000} = \frac{5}{8}$$

STEP 1: Rewrite 62.5% as a decimal: .625

STEP 2: Rewrite .625 as a fraction: $\frac{625}{1000}$

STEP 3: To simplify $\frac{625}{1000}$ divide the numerator and the denominator by 125. 62.5% is the same as $\frac{5}{8}$.

EXERCISE 30b

Rewrite each percent as a fraction or a mixed number in lowest terms.

1. 12.5%   2. .5%   3. 37.5%   4. 6.25%

5. 87.5%   6. .2%   7. 22.5%   8. 32.5%

9. 112.5%   10. 337.5%   11. 2.5%   12. .04%

> **WORD PROBLEM**
>
> This word problem gives you a percent but asks you express the answer as a fraction.
>
> During a snowstorm only 37.5% of the employees at a certain plant came to work. What fraction of the employees came to work?

Check your answers on page 235.

Rewriting Percents with Fractions as Fractions

Sometimes percents include fractions instead of decimals. For example, the following percents mean the same thing:

$$12.5\% \qquad 12\tfrac{1}{2}\%$$

Sometimes you need to rewrite a percent that includes a fraction as a fraction. To do that, first rewrite the fraction in the percent as a decimal.

Example 4: Rewrite $37\tfrac{1}{2}\%$ as a fraction.

Step 1	Step 2	Step 3
$37\tfrac{1}{2}\% = 37.5\%$	$37.5\% = .375$	$.375 = \dfrac{375}{1000} = \dfrac{3}{8}$

STEP 1: Rewrite the fraction $37\tfrac{1}{2}\%$ as a decimal. The percent becomes 37.5%.
STEP 2: Rewrite the percent as a decimal.
STEP 3: Rewrite the decimal as a fraction and simplify.

EXERCISE 30c

Rewrite each percent as a fraction in lowest terms.

1. $112\tfrac{1}{2}\%$
2. $\tfrac{1}{2}\%$
3. $37\tfrac{1}{2}\%$
4. $6\tfrac{1}{4}\%$

5. $87\tfrac{1}{2}\%$
6. $\tfrac{1}{5}\%$
7. $\tfrac{1}{10}\%$
8. $\tfrac{3}{4}\%$

9. $7\tfrac{3}{4}\%$
10. $\tfrac{3}{10}\%$
11. $6\tfrac{3}{5}\%$
12. $99\tfrac{9}{10}\%$

WORD PROBLEM

This word problem asks you to express a percent as a fraction.

At the plant one day, $8\frac{1}{2}\%$ of the goods produced were defective. What fraction of the goods were defective?

Check your answers on page 235.

Lesson 31

Rewriting Fractions as Percents

To solve some problems it is necessary to rewrite a fraction as a percent. For example, you may use fractions when you work a problem but use a percent to express the answer.

Rewriting Fractions with Denominators of 100 as Percents

When a fraction has a denominator of 100, change it to a percent by dropping its denominator and writing a percent sign after its numerator.

Example 1: Rewrite $\frac{23}{100}$ as a percent.

       Step 1   **Step 2**

       $\frac{23}{100}$ $\frac{23}{\cancel{100}}\% = 23\%$

STEP 1: Write the fraction.

STEP 2: Drop the denominator and the bar that separates it from the numerator. Write a percent sign following the numerator. $\frac{23}{100}$ equals 23%.

EXERCISE 31a

Rewrite each fraction as a percent.

1. $\frac{1}{100}$ 2. $\frac{17}{100}$ 3. $\frac{33}{100}$ 4. $\frac{51}{100}$ 5. $\frac{99}{100}$

WORD PROBLEM

This word problem asks you to express a fraction as a percent.

To plan a woodworking job, Alan had to shade $\frac{33}{100}$ of a grid, as shown by the illustration. What percent of the grid did he have to shade?

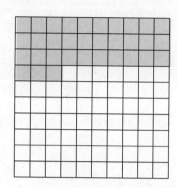

Shaded portion: $\frac{33}{100}$

Check your answers on page 235.

Rewriting Other Fractions as Percents

To rewrite a fraction with a denominator other than 100 as a percent, first rewrite the fraction as a decimal. Then rewrite the decimal as a percent.

Example 2: Rewrite $\frac{5}{8}$ as a percent.

Step 1	Step 2	Step 3
$\frac{5}{8}$	$\begin{array}{r} .625 \\ 8\overline{)5.000} \\ \underline{4\,8} \\ 20 \\ \underline{16} \\ 40 \\ \underline{40} \\ 0 \end{array}$	$.62.5 = 62.5\%$

STEP 1: Write the fraction.

STEP 2: To rewrite the fraction as a decimal, divide 5 (the numerator) by 8 (the denominator). The result is .625.

STEP 3: Rewrite .625 as a percent by moving its decimal point two places to the right and adding a percent sign. $\frac{5}{8}$ equals 62.5%.

When you divide some fractions to rewrite them as decimals, the quotient will not come out even. In such cases, divide to the hundredths place of the quotient and express the remainder as a fraction with the divisor as the denominator.

Chapter 1: Percents, Decimals, and Fractions

Example 3: Rewrite $\frac{1}{6}$ as a percent.

Step 1	Step 2	Step 3
$\frac{1}{6}$	$6 \overline{)1.00}^{.16\frac{4}{6} = .16\frac{2}{3}}$ $\phantom{6\overline{)}}\underline{6}$ $\phantom{6\overline{)1}}40$ $\phantom{6\overline{)1}}\underline{36}$ $\phantom{6\overline{)1.0}}4$	$.16\frac{2}{3} = 16\frac{2}{3}\%$

STEP 1: Write the fraction.

STEP 2: Rewrite the fraction as a decimal. Because the quotient will never come out even, divide to the hundredths place of the quotient: .16. Write the remainder over the divisor: $\frac{4}{6}$. Simplify the answer: $.16\frac{4}{6} = .16\frac{2}{3}$.

STEP 3: Rewrite $.16\frac{2}{3}$ as a percent. $\frac{1}{6}$ equals $16\frac{2}{3}\%$.

EXERCISE 31b

Rewrite each fraction as a percent.

1. $\frac{1}{2}$
2. $\frac{1}{16}$
3. $\frac{1}{8}$
4. $\frac{1}{20}$

5. $\frac{1}{4}$
6. $\frac{3}{4}$
7. $\frac{4}{5}$
8. $\frac{3}{8}$

9. $\frac{3}{16}$
10. $\frac{2}{5}$
11. $\frac{7}{25}$
12. $\frac{5}{16}$

13. $\frac{5}{6}$
14. $\frac{1}{3}$
15. $\frac{2}{3}$
16. $\frac{3}{7}$

> **WORD PROBLEM**
>
> This word problem asks you to express a fraction as a percent.
>
> Of all the households on Franklin Street, $\frac{15}{16}$ belong to the community association. Correct to the nearest hundredth, what percent of the households belong to the association?

Check your answers on page 235.

Recognizing Common Equivalents

The equivalents for some percents are used more often than others. It is worthwhile to take the time to memorize the common equivalents. That way, when you need to use them, you won't have to take the time to rewrite one kind of number as another. Memorize the following equivalents.

COMMON EQUIVALENTS

	Fraction	Decimal	Percent
Tenths	$\frac{1}{10}$	.1	10%
	$\frac{3}{10}$	.3	30%
	$\frac{7}{10}$	.7	70%
	$\frac{9}{10}$	.9	90%
Eighths	$\frac{1}{8}$	.125	$12\frac{1}{2}$% or 12.5%
	$\frac{3}{8}$	.375	$37\frac{1}{2}$% or 37.5%
	$\frac{5}{8}$	.625	$62\frac{1}{2}$% or 62.5%
	$\frac{7}{8}$	.875	$87\frac{1}{2}$% or 87.5%
Sixths	$\frac{1}{6}$	$.16\frac{2}{3}$	$16\frac{2}{3}$%
	$\frac{5}{6}$	$.83\frac{1}{3}$	$83\frac{1}{3}$%
Fifths	$\frac{1}{5}$	.2	20%
	$\frac{2}{5}$	.4	40%
	$\frac{3}{5}$	.6	60%
	$\frac{4}{5}$	.8	80%
Quarters	$\frac{1}{4}$	.25	25%
	$\frac{3}{4}$	.75	75%
Thirds	$\frac{1}{3}$	$.33\frac{1}{3}$	$33\frac{1}{3}$%
	$\frac{2}{3}$	$.66\frac{2}{3}$	$66\frac{2}{3}$%
Half	$\frac{1}{2}$	.5	50%

EXERCISE 32

Fill in the blanks. Try to do the exercise from memory.

	Percent	Fraction	Decimal
1.	60%	_____	_____
2.	_____	_____	.25
3.	_____	$\frac{1}{2}$	_____
4.	75%	_____	_____
5.	_____	$\frac{1}{10}$	_____
6.	_____	_____	.2
7.	30%	_____	_____
8.	_____	_____	$.16\frac{2}{3}$
9.	62.5%	_____	_____
10.	_____	$\frac{1}{3}$	_____
11.	_____	$\frac{1}{8}$	_____
12.	_____	_____	$.66\frac{2}{3}$

WORD PROBLEM

This word problem asks you to express a fraction as a percent.

Miguel is buying a stereo on an installment plan. He paid $\frac{1}{3}$ of the price as a down payment. What percent of the price was the down payment?

Check your answers on page 235.

Chapter 2 SOLVING PERCENT PROBLEMS

In the four lessons in this chapter you will learn to use both a percent triangle and formulas to solve the three basic kinds of percent problems. The lessons cover some of the skills you will use in the next chapter in this unit.

Lesson 33 Understanding the Percent Triangle

In Lessons 34–36 you will learn to solve the three basic types of percent problems. Before you study those lessons, you need to know something about the numbers in a percent problem and how they are multiplied or divided to solve problems.

Identifying the Numbers in a Percent Problem

Every percent problem is made up of three numbers: the **percent**, the **whole**, and the **part**. The following percent statement has its three numbers labelled:

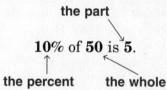

Percent problems like the following three examples give you two of those numbers and ask you to find the third:

What is 10% of 50? You must find **the part**.
What percent of 50 is 5? You must find **the percent**.
5 is 10% of **what number**? You must find **the whole**.

The word *of* is used in many percent problems. Usually the number that follows *of* is the whole, as in the examples above.

EXERCISE 33a

Complete all three statements about each problem. The first one is done for you as an example.

1. What is 15% of 70?
   (a) 15% is the ___percent___ .
   (b) 70 is the ___whole___ .
   (c) You must find the ___part___ .

2. What percent of 90 is 4.5?
   (a) 90 is the _____ .
   (b) 4.5 is the _____ .
   (c) You must find the _____ .

3. 18 is 20% of what number?
   (a) 18 is the _____ .
   (b) 20% is the _____ .
   (c) You must find the _____ .

4. 9 is what percent of 4.5?
   (a) 9 is the _____ .
   (b) 4.5 is the _____ .
   (c) You must find the _____ .

5. 2 is 25% of what number?
   (a) 2 is the _____ .
   (b) 25% is the _____ .
   (c) You must find the _____ .

6. 70% of 15 is what?
   (a) 70% is the _____ .
   (b) 15 is the _____ .
   (c) You must find the _____ .

WORD PROBLEM

Tell whether the problem asks you to find the percent, the whole, or the part.

Because he correctly answered 15 of the questions on his exam for a driver's license, Harold's score was 75%. How many questions were on the exam?

Check your answers on page 235.

Using the Percent Triangle

The method for finding each unknown number is different, so it is useful to have a handy tool for remembering which way to solve each type of problem. Many people use a percent triangle as that handy tool.

PERCENT TRIANGLE

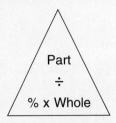

132 UNIT 2: Percents

To use the percent triangle, cover the word that represents the number you want to find. The math operation you need is shown in the uncovered section of the triangle.

Example: What math operation would you use to solve this problem: Find 8.25% of $1.29.

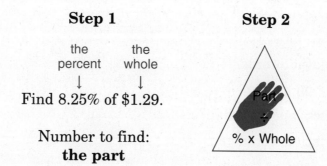

Step 1

    the   the
  percent whole
     ↓   ↓
Find 8.25% of $1.29.

Number to find:
the part

Step 2

STEP 1: Determine what number the problem asks you to find. Because the percent and the whole are given, you are asked to find the part.

STEP 2: In the percent triangle cover "part" and the division sign. The uncovered section of the triangle shows how to find the part. You would multiply the percent times the whole: 8.25% × $1.29.

When you cover "percent" in the triangle, the uncovered section looks like this. It shows that to find the percent you would divide the part by the whole.

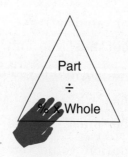

When you cover "whole" in the triangle, the uncovered section looks like this. It shows that to find the whole you would divide the part by the percent.

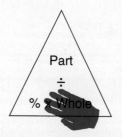

Chapter 2: Solving Percent Problems   **133**

EXERCISE 33b

With the help of the percent triangle, answer both questions about each problem. The first one is done for you as an example.

1. What is 15% of 70?
   (a) What number must you find? ____the part____
   (b) What calculation would you do? ____Multiply 15% times 70.____

2. What percent of 90 is 4.5?
   (a) What number must you find? _____
   (b) What calculation would you do? _____

3. 18 is 20% of what number?
   (a) What number must you find? _____
   (b) What calculation would you do? _____

4. 9 is what percent of 4.5?
   (a) What number must you find? _____
   (b) What calculation would you do? _____

5. 2 is 25% of what number?
   (a) What number must you find? _____
   (b) What calculation would you do? _____

6. 70% of 15 is what?
   (a) What number must you find? _____
   (b) What calculation would you do? _____

> **WORD PROBLEM**
>
> What calculation would you do to solve the following problem?
>
> Because he correctly answered 15 of the questions on his exam for a driver's license, Harold's score was 75%. How many questions were on the exam?

Check your answers on page 235.

134   UNIT 2: Percents

Lesson 34: Finding the Part

In Lesson 33 you used the percent triangle to find the correct calculation for solving any percent problem.

The percent triangle shows that to find the part you multiply the percent times the whole.

The method the percent triangle shows for finding the part can be expressed in a formula:

$$\text{part} = \text{percent} \times \text{whole}$$

A shorter form of the same formula is

$$\text{part} = \% \times \text{whole}$$

To find the part in a percent problem, you don't actually multiply by the percent itself. Rather, you rewrite the percent as an equivalent decimal or fraction—whichever is easier to work with—before you multiply.

Finding the Part by Using the Decimal Equivalent of a Percent

When a percent is not listed among the equivalents on page 129, substitute its decimal equivalent in the formula for finding the part. You learned to rewrite percents as decimals in Lesson 28.

Example 1: What is 51% of 250?

Step 1	**Step 2**	**Step 3**
part = % × whole	part = % × whole	part = % × whole
	part = .51 × 250	part = .51 × 250
		part = 127.5

STEP 1: Write the formula for finding the part.

STEP 2: Rewrite the formula, substituting the numbers from the problem. Instead of writing 51% in the formula, use .51, its decimal equivalent.

STEP 3: Multiply to solve the problem 127.5 is 51% of 250.

Chapter 2: Solving Percent Problems   135

EXERCISE 34a

Solve each problem. Round money answers to the nearest cent.

1. What is 16% of 150?
2. What is .7% of 200?
3. What is 8% of 350?
4. What is 9% of 400?
5. What is 18% of $72?
6. What is 1.5% of 245?
7. What is 11% of $34?
8. What is .14% of $56?
9. What is 62% of $11?
10. What is 9% of $6?
11. What is 12% of 72?
12. What is 7.2% of $143?
13. What is 115% of 240?
14. What is 37.8% of $1500?
15. What is 2.33% of 46?
16. What is .142% of $390?
17. What is 212% of $9?
18. What is 10.5% of $24.80?
19. What is 108.75% of $20?
20. What is 8.75% of $25?

WORD PROBLEM

The following percent problem asks you to find the part.

Simone saves 12% of her pay. She earns $1475 per month. How much does she save each month?

Check your answers on page 236.

Finding the Part by Using the Fraction Equivalent of a Percent

If you wish, you may substitute a fraction for the percent in the formula for finding the part. (See the table of Common Equivalents on page 129 and, if necessary, review Lesson 30.)

Example 2: What is $33\frac{1}{3}$% of $150?

Step 1	Step 2	Step 3
part = % × whole	part = % × whole	part = % × whole
	part = $\frac{1}{3}$ × $150	part = $\frac{1}{3}$ × $150
		part = $50

UNIT 2: Percents

STEP 1: Write the formula for finding the part.

STEP 2: Rewrite the formula, substituting the numbers from the problem. Instead of writing $33\frac{1}{3}\%$ in the formula, use $\frac{1}{3}$, its fraction equivalent.

STEP 3: Multiply to solve the problem: $50 is $33\frac{1}{3}\%$ of $150.

When a percent is larger than 100%, you may rewrite it as a mixed number. For example, you can rewrite 125% as a mixed number because it is made up of 100% + 25%. You know that 100% equals 1 and that the fraction equivalent of 25% is $\frac{1}{4}$. Therefore, 125% equals $1\frac{1}{4}$.

EXERCISE 34b

Solve each problem. Use mixed numbers to solve the problems with percents larger than 100%.

1. What is 60% of 80?
2. What is 50% of 128?
3. What is 70% of 50?
4. What is 225% of 160?
5. What is 75% of 600?
6. What is $66\frac{2}{3}\%$ of 126?
7. What is $87\frac{1}{2}\%$ of 160?
8. What is $16\frac{2}{3}\%$ of 360?
9. What is $112\frac{1}{2}\%$ of $98?
10. What is $37\frac{1}{2}\%$ of $196?
11. What is $83\frac{1}{3}\%$ of $72?
12. What is 110% of $12.80?
13. What is 51% of 1000?
14. What is 13% of $10?

WORD PROBLEM

The following percent problem asks you to find the part.

Luz teaches a cooking class of 45 people. $33\frac{1}{3}\%$ of the class are men. How many men are in the class?

Check your answers on page 236.

Lesson 35

Finding the Percent

As you know, you can use the percent triangle to find out which operation to use to solve a particular percent problem.

The percent triangle shows that to find a percent you divide the part by the whole.

The method for finding a percent can be expressed in a formula:

$$\text{percent} = \frac{\text{part}}{\text{whole}}$$

A shorter form of the same formula is

$$\% = \frac{\text{part}}{\text{whole}}$$

As you will see, the last step in using this formula is always to rewrite a fraction or a decimal as a percent.

Finding the Percent by Dividing

One way to use the formula to find the percent is to divide the part by the whole. You will need to rewrite the decimal in your answer as a percent. (You learned to do this in Lesson 29.)

Example 1: What percent of 150 is 39?

Step 1

$$\% = \frac{\text{part}}{\text{whole}}$$

Step 2

$$\% = \frac{\text{part}}{\text{whole}}$$

$$\% = \frac{39}{150}$$

Step 3

$$\% = \frac{\text{part}}{\text{whole}}$$

$$\% = 39 \div 150$$

$$\% = .26$$

$$.26 = 26\%$$

STEP 1: Write the formula for finding the part.

STEP 2: Rewrite the formula, substituting the numbers from the problem.

STEP 3: Divide to solve the problem. Rewrite the decimal as a percent.
39 is 26% of 150.

EXERCISE 35a

Find the following percents by dividing.

1. What percent of 75 is 9?
2. What percent of 200 is 16?
3. What percent of 120 is 102?
4. What percent of 200 is 3?
5. What percent of 500 is 24?
6. 388 is what percent of 400?
7. 4.5 is what percent of 150?
8. .3 is what percent of 6?
9. 7 is what percent of 400?
10. 40 is what percent of 320?

WORD PROBLEM

The following problem asks you to find the percent.

Manny Ortega had 300 copies of a popular new book to sell in his store. By the end of the first week he had sold 264 copies. What percent of the books did Manny sell that week?

Check your answers on page 236.

Finding the Percent by Simplifying a Fraction

Another way to use the formula to find a percent is to simplify the fraction in the formula. Often the resulting fraction will be one whose equivalent percent you know.

Example 2: 10 is what percent of 12?

Step 1

$$\% = \frac{\text{part}}{\text{whole}}$$

Step 2

$$\% = \frac{\text{part}}{\text{whole}}$$
$$\% = \frac{10}{12}$$

Step 3

$$\% = \frac{\text{part}}{\text{whole}}$$
$$\% = \frac{10}{12}$$
$$\% = \frac{5}{6}$$

$$\frac{5}{6} = 83\frac{1}{3}\%$$

STEP 1: Write the formula for finding the part.

STEP 2: Rewrite the formula, substituting the numbers from the problem.

STEP 3: Simplify the fraction and rewrite it as an equivalent percent.
10 is $83\frac{1}{3}\%$ of 12.

When you use this method to find a percent, the fraction that results when you simplify in Step 3 may not be one whose equivalent you know. If you do not know the equivalent percent, divide the numerator by the denominator, as in Example 1.

EXERCISE 35b

Find the following percents by simplifying fractions.

1. What percent of 40 is 30?
2. What percent of 16 is 8?
3. What percent of 70 is 7?
4. What percent of 20 is 4?
5. What percent of 25 is 25?
6. 30 is what percent of 90?
7. 25 is what percent of 200?
8. 50 is what percent of 75?
9. 20 is what percent of 32?
10. 9 is what percent of 18?
11. 13 is what percent of 78?
12. 40 is what percent of 48?
13. What percent of 200 is 98?
14. What percent of 300 is 39?
15. What percent of 1000 is 70?
16. What percent of 4000 is 250?

WORD PROBLEM

The following problem asks you to find the percent.

Jesus Guzman bought a used car for $1500. He made a $300 down payment. What percent of the purchase price was the down payment?

Check your answers on page 236.

Finding Percents Larger than 100%

When the part is larger than the whole, the percent will be larger than 100%. (Remember that the whole often follows the word *of*.)

Example 3: 32 is what percent of 20?

Step 1

$$\% = \frac{\text{part}}{\text{whole}}$$

Step 2

$$\% = \frac{\text{part}}{\text{whole}}$$
$$\% = \frac{32}{20}$$

Step 3

$$\% = \frac{\text{part}}{\text{whole}}$$
$$\% = \frac{32}{20}$$
$$\% = 1\frac{3}{5}$$

$$1\frac{3}{5} = 160\%$$

STEP 1: Write the formula for finding the part.

STEP 2: Rewrite the formula, substituting the numbers from the problem.

STEP 3: Simplify the fraction and rewrite the mixed number as an equivalent percent. Since $\frac{3}{5}$ is equivalent to 60%, $1\frac{3}{5}$ equals 160%.

Example 3 was solved by simplifying fractions. It could also be solved by dividing, as can any problem in which you are looking for the percent.

EXERCISE 35c

Find the following percents.

1. What percent of 8 is 24?
2. What percent of 1.25 is 25?
3. What percent of 3 is 6?
4. What percent of 4 is 12?

5. What percent of 4 is 14?
6. 2 is what percent of 1?
7. 156 is what percent of 4?
8. 6 is what percent of 4?
9. 75 is what percent of 50?
10. 36 is what percent of 27?
11. 88 is what percent of 55?
12. 75 is what percent of 12.5?

WORD PROBLEM

The following problem asks you to find a percent larger than 100%.

Receipts from Lupe's Lunch Pail were $480 last Saturday. This Saturday the receipts were $600. What percent of last Saturday's receipts were this week's receipts?

Check your answers on page 236.

Lesson 36

Finding the Whole

Some percent problems give you the part and the percent and ask you to find the whole.

The percent triangle shows that to find the whole you divide the part by the percent.

The method for finding the whole is expressed in this formula:

$$\text{whole} = \frac{\text{part}}{\text{percent}}$$

A shorter form of the same formula is

$$\text{whole} = \frac{\text{part}}{\%}$$

To divide by the percent in the formula, you need to rewrite it as a decimal or a fraction.

142 UNIT 2: Percents

Dividing by a Decimal to Find the Whole

To find the whole you can always rewrite the percent as a decimal and divide.

Example 1: 84 is 21% of what number?

Step 1	Step 2	Step 3
whole = $\frac{\text{part}}{\%}$	whole = $\frac{\text{part}}{\%}$ whole = 84 ÷ .21	whole = $\frac{\text{part}}{\%}$ whole = 84 ÷ .21 whole = 400

STEP 1: Write the formula for finding the whole.

STEP 2: Rewrite the formula, substituting the numbers from the problem. Instead of writing 21% in the formula, use .21, its decimal equivalent.

STEP 3: Divide to solve the problem. 84 is 21% of 400.

EXERCISE 36a

In each of the following problems, divide by the decimal equivalent of the percent to find the whole.

1. 840 is .25% of what number?
2. 5 is 2% of what number?
3. 10 is 1% of what number?
4. 34.5 is 23% of what number?
5. 8 is 4% of what number?
6. 7% of what number is 21?
7. 3% of what number is 15?
8. 47% of what amount is $11.75?
9. 8.75% of what amount is $78.75?
10. $59.50 is 119% of what amount?
11. $7.20 is 128% of what amount?
12. .019 is 38% of what number?
13. .42 is 21% of what number?
14. 1.6 is 48% of what number?
15. 99.44 is 99.44% of what number?
16. $210 is 105% of what amount?

> **WORD PROBLEM**
>
> The following problem asks you to find the whole.
>
> Kenia's $3840 bonus this year amounted to 16% of her regular salary. How much is her salary?

Check your answers on page 237.

Dividing by a Fraction to Find the Whole

When you know the fraction equivalent of the percent in a problem, substitute it for the percent in the formula.

Example 2: 16 is 25% of what number?

Step 1

$$\text{whole} = \frac{\text{part}}{\%}$$

Step 2

$$\text{whole} = \frac{\text{part}}{\%}$$
$$\text{whole} = 16 \div \frac{1}{4}$$

Step 3

$$\text{whole} = \frac{\text{part}}{\%}$$
$$\text{whole} = 16 \times \frac{4}{1}$$
$$\text{whole} = 64$$

STEP 1: Write the formula for finding the whole.

STEP 2: Rewrite the formula, substituting the numbers from the problem. Instead of writing 25% in the formula, use $\frac{1}{4}$, its fraction equivalent.

STEP 3: Divide to solve the problem. (Remember that when you divide by a fraction, you must invert it and multiply.) 16 is 25% of 64.

EXERCISE 36b

In each of the following problems, divide by the fraction equivalent of the percent to find the whole.

1. 8 is 25% of what number?
2. 2 is $12\frac{1}{2}$% of what number?
3. 36 is 30% of what number?
4. 15 is $33\frac{1}{3}$% of what number?
5. 48 is $37\frac{1}{2}$% of what number?
6. $87\frac{1}{2}$% of what number is 49?
7. $12\frac{1}{2}$% of what number is 8?
8. 40% of what number is 64?
9. $66\frac{2}{3}$% of what number is 48?
10. 125% of what amount is $50?
11. 500 is 200% of what amount?
12. 12 is 120% of what number?
13. 240 is 150% of what number?
14. 840 is $133\frac{1}{3}$% of what number?
15. 100 is 250% of what number?
16. 78 is 39% of what number?
17. 57 is 19% of what number?
18. 51 is 51% of what number?

WORD PROBLEM

The following problem asks you to find the whole.

Sixty-three of the merchants on Pine Avenue belong to the Chamber of Commerce. They make up 70% of the Pine Avenue merchants. How many merchants are there on Pine Avenue altogether?

Check your answers on page 237.

PERCENT SKILLS REVIEW

Part A. Complete the statements about the following figures.

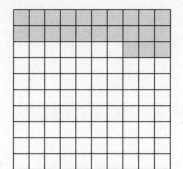

Figure 1

1. _____ % of Figure 1 is shaded.

2. _____ % of Figure 1 is not shaded.

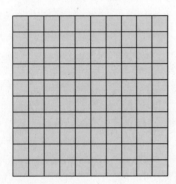 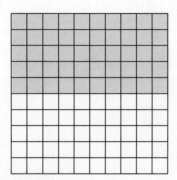

Figure 2

3. In Figure 2, _____ % of one large square is shaded.

Figure 3

4. Figure 3 shows _____ % of a dollar.

Part B. Rewrite each number as a percent.

1. 2.4   2. $\frac{5}{8}$ 3. $\frac{2}{3}$ 4. $.37\frac{1}{2}$

5. .006   6. .333   7. $\frac{1}{20}$ 8. $\frac{5}{6}$

9. 70.065   10. .299

Part C. Rewrite each percent as a decimal or a mixed decimal.

1. 3.2%
2. 42%
3. .06%
4. 5%
5. 109%
6. $12\frac{1}{2}\%$
7. $16\frac{2}{3}\%$
8. 20%
9. $333\frac{1}{3}\%$
10. 50%

Part D. Rewrite each percent as a fraction or a mixed number in simplest form.

1. $87\frac{1}{2}\%$
2. $83\frac{1}{3}\%$
3. 112.5%
4. 60%
5. 75%
6. $33\frac{1}{3}\%$
7. 50%
8. 430%
9. 1%
10. 120%

Part E. Solve each problem.

1. What is 75% of 80?
2. What percent of 28 is 7?
3. What is .5% of $92?
4. What percent of $8 is $80?
5. What is $66\frac{2}{3}\%$ of 117?
6. What is 220% of 500?
7. 120 is 150% of what number?
8. What is 12% of 30?
9. 200% of what number is 206?
10. What is 1.35% of 150?
11. 16 is what percent of 25.6?
12. 9% of what number is 27?
13. 12 is 1.5% of what number?
14. What is $.12\frac{1}{2}\%$ of 12,800?
15. What percent of 45 is 90?
16. What percent of $90 is $45?
17. What is 100% of 9.8?
18. 20 is what percent of 32?
19. $35 is 70% of what amount?
20. What percent of 48 is 8?
21. $37\frac{1}{2}\%$ of what number is 6?
22. $16 is what percent of $5?
23. $33\frac{1}{3}\%$ of what number is 45?
24. What is 1000% of .075?

Check your answers on page 237.

Skills Review

Chapter 3

WORKING WITH PERCENTS

The first lesson in this chapter teaches you how to estimate solutions to percent problems. The other two lessons show how to solve one-step and multistep percent problems similar to the percent problems on the GED.

Estimating Solutions to Percent Problems

It is often useful to estimate the solution to a percent problem. Sometimes you may need to know only an approximate answer, not an exact one.

For example, your employer may tell you that your medical insurance will cost you about 25% more this year than last. You know that last year you paid $389. You want to know roughly how much more you will pay this year. Instead of working the problem out you can estimate. You know that 25% is $\frac{1}{4}$. You also know that $389 is about $400. You can estimate that your medical insurance will cost about $100 more this year because $\frac{1}{4}$ of $400 is $100.

Estimating the Part by Changing the Whole

In Lesson 34, you found the part in percent problems. You multiplied the whole by the percent:

$$\text{part} = \% \times \text{whole}$$

Using the fraction equivalent of a percent, you can estimate the part. Just change the whole to a number that is easy to work with.

Example 1: What number is about $16\frac{2}{3}\%$ of 281?

Step 1	Step 2	Step 3
$16\frac{2}{3}\% = \frac{1}{6}$	281 can be changed to 300.	$\frac{1}{6} \times 300 = 50$

STEP 1: Rewrite $16\frac{2}{3}\%$ as its fraction equivalent, $\frac{1}{6}$.

STEP 2: Change 281 to 300, the closest round number that can be multiplied by $\frac{1}{6}$ easily.

STEP 3: $\frac{1}{6}$ of 300 is 50. Therefore, $16\frac{2}{3}\%$ of 281 is about 50. (The exact answer is $46\frac{2}{3}$.)

148 UNIT 2: Percents

Changing the whole in problems like this takes some practice. You should change it to a number that will multiply evenly by the fraction.

It is not always helpful to use round numbers when you estimate. For example, when you multiply by $\frac{1}{3}$, it is best to change the whole number to a multiple of three.

You can also estimate when you multiply by the decimal equivalent of a percent.

Example 2: What number is about 20% of 138?

Step 1	**Step 2**	**Step 3**
20% = .2	138 can be changed to 140.	.2 × 140 = 28

STEP 1: Rewrite 20% as .2, its decimal equivalent.

STEP 2: Change 138 to 140, the closest number that can be multiplied by .2 easily.

STEP 3: 140 × .2 = 28. Therefore, 20% of 138 is about 28. (The exact answer is 27.6.)

When you estimate using a decimal, change the whole to a number that will multiply easily by the decimal.

EXERCISE 37a

In each of the following problems, estimate the part. (Change the whole before multiplying.)

1. What number is about 50% of 206?
2. What number is about $33\frac{1}{3}$% of 38?
3. What number is about 25% of 45?
4. What number is about $87\frac{1}{2}$% of 98?
5. What number is about 40% of 322?
6. About how much is $12\frac{1}{2}$% of 325?
7. About how much is 70% of 231?
8. About how much is $37\frac{1}{2}$% of 65?
9. About how much is $16\frac{2}{3}$% of 38?
10. About how much is 5% of 103?

> **WORD PROBLEM**
>
> In the following problem, estimate the part. (Change the whole before multiplying.)
>
> Evita spends $33\frac{1}{3}\%$ of her income on rent and utilities. Her annual income last year was $17,897. About how much did Evita spend on rent and utilities last year?

Check your answers on page 238.

Estimating the Part by Changing the Percent

You can estimate the part another way. Rewrite the percent as one that has an equivalent fraction or decimal that is easy to work with.

Example 3: What number is about 48% of 200?

Step 1
48% can be changed to 50%, which equals $\frac{1}{2}$.

Step 2
$\frac{1}{2} \times 200 = 100$

STEP 1: Change 48% to 50%, the closest percent with a fraction equivalent that can be multiplied by 200 easily.

STEP 2: $\frac{1}{2} \times 200 = 100$. Therefore, 48% of 200 is about 100. (The exact answer is 96.)

EXERCISE 37b

In each of the following problems, estimate the part. (Change the percent before multiplying.)

1. About how much is 74% of 400?
2. About how much is 51% of 150?
3. About how much is 9% of 300?
4. About how much is 47% of 86?
5. About how much is 33% of 999?
6. What number is about 26% of 80?
7. What number is about 34% of 18?
8. What number is about 13% of 16?
9. What number is about 17% of 72?
10. What number is about 39% of 50?

> **WORD PROBLEM**
>
> In the following problem, estimate the part. (Change the percent before multiplying.)
>
> In a recent year 47% of the 800 applicants for jobs as police officers failed the physical examination. About how many applicants failed the physical?

Check your answers on page 238.

Estimating the Whole

In Lesson 36 you found the whole in percent problems. You divided the part by the percent:

$$\text{whole} = \frac{\text{part}}{\%}$$

When a percent problem asks you to find the whole, you can estimate the answer. You can change either the part or the percent so that the numbers in the problem are easy to work with. In the following example, the percent is changed.

Example 4: 20 is about 49% of what number?

Step 1
49% can be changed to 50%, which equals $\frac{1}{2}$.

Step 2
$20 \div \frac{1}{2} = 40$

STEP 1: Change 49% to 50% because $\frac{1}{2}$, its fraction equivalent, can be divided into 20 easily.

STEP 2: $20 \div \frac{1}{2} = 40$. Therefore, 49% of 20 is about 40. (Rounded to the nearest hundredth, the whole is actually 40.82.)

In some problems it is better to change the part, not the percent. For example, to estimate the whole when 19 is 25% of it, divide 20 by $\frac{1}{4}$.

Chapter 3: Working with Percents

EXERCISE 37c

In each of the following problems, estimate the whole. (Change either the percent or the part before dividing.)

1. 15 is about 32% of what number?
2. 20 is about 27% of what number?
3. 29 is about $33\frac{1}{3}$% of what number?
4. 6.2 is about 11% of what number?
5. 4.1 is about $12\frac{1}{2}$% of what number?
6. 8 is about 24% of what number?
7. 116 is about 9% of what number?
8. 362 is about 90% of what number?
9. 20 is about 19% of what number?
10. 1.7 is about $16\frac{2}{3}$% of what number?

WORD PROBLEM

In the following problem, estimate the whole. (Change either the percent or the part before dividing.)

At Arvida's Shopping Mall there are 40 part-time employees. They make up about 24.7% of the employees at the mall. About how many employees are there at the mall?

Check your answers on page 238.

Estimating the Percent

In Lesson 35 you found the percent when the whole and the part were given. You divided the part by the whole:

$$\% = \frac{\text{part}}{\text{whole}}$$

When a problem asks you to find the percent, you can often estimate the answer by simplifying a fraction. Change either the part or the whole so the fraction they make (part/whole) can be simplified. Write the fraction as a percent to give the estimated answer. In the following example the whole is changed.

Example 5: 15 is about what percent of 29?

Step 1
29 can be changed to 30.

Step 2
$\frac{15}{30} = \frac{1}{2} = 50\%$

STEP 1: Change 29 to 30 because a fraction with 15 and 30 can be simplified.
STEP 2: $\frac{15}{30} = \frac{1}{2}$. Therefore, 15 is about 50% of 29.

In some problems you may want to change the part rather than the whole. For example, if you want to estimate what percent 23 is of 250, it would probably be easiest to change 23 to 25. Then the fraction is $\frac{25}{250}$, which simplifies to $\frac{1}{10}$. It can be rewritten as 10%.

EXERCISE 37d

In each of the following problems, estimate the percent. (Change either the whole or the part and simplify the fraction they make.)

1. What percent of 40 is 29?
2. What percent of 16 is 11?
3. What percent of 69 is 7?
4. What percent of 21 is 4?
5. What percent of 250 is 249?
6. 30 is what percent of 89?
7. 21 is what percent of 200?
8. 23 is what percent of 75?
9. 21 is what percent of 32?
10. 9 is what percent of 19?

> **WORD PROBLEM**
>
> In the following example, estimate the percent. (Change either the whole or the part and simplify the fraction they make.)
>
> John found that 12 of his 61 customers on Monday paid for their purchases with food stamps. About what percent of his customers used food stamps?

Check your answers on page 238.

Solving One-Step Percent Word Problems

The steps for solving percent word problems are similar to those for solving other word problems:

- Restate the problem.
- Choose the formula to use.
- Estimate the solution.
- Solve the problem.

As an example, these four steps are applied to each of three problems.

Chapter 3: Working with Percents   153

Example: Problem A: The sales tax in Rico's city is 5% of an item's selling price. How much tax did Rico pay on a $43 ladder?

Problem B: During a sale, UpTown Clothiers took $18 off the price of an $80 coat. By what percent was the coat marked down?

Problem C: Zarena's 9.5% rent increase added $38 to her monthly rent. What was Zarena's monthly rent before the increase?

STEP 1: Restate the problem.

To solve percent problems, three numbers are important: the part, the percent, and the whole. One-step word problems give you two of those numbers and ask you to find the third. When you read a problem, identify which two numbers it gives. Two clues can help you:

The percent includes a percent sign (%) or the word *percent*.
The whole often follows the word *of*.

Restating the problem in a very simple form can also help you decide which numbers the problem gives and which one you must find. The example problems are restated below:

Problem A
The sales tax in Rico's city is 5% of an item's selling price. How much tax did Rico pay on a $43 ladder?

The problem can be restated simply:

What is 5% of $43?

Problem B
During a sale, UpTown Clothiers took $18 off the price of an $80 coat. By what percent was the coat marked down?

The problem can be restated simply:

$18 is what percent of $80?

Problem C
Zarena's 9.5% rent increase added $38 to her monthly rent. What was Zarena's monthly rent before the increase?

The problem can be restated simply:

$38 is 9.5% of what amount?

STEP 2: Choose the formula to use.

To solve a percent problem, you need to use the correct formula. Inserting the numbers from a problem in a percent triangle can help you find the formula you need.

When you have the correct formula, rewrite it substituting the numbers from the problem. Look how that can be done with the example problems:

Problem A
The sales tax in Rico's city is 5% of an item's selling price. How much tax did Rico pay on a $43 ladder?

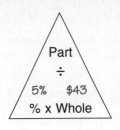

part = % × whole
part = 5% × $43

Problem B
During a sale, UpTown Clothiers took $18 off the price of an $80 coat. By what percent was the coat marked down?

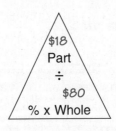

% = part/whole
% = $18/$80

Problem C
Zarena's 9.5% rent increase added $38 to her monthly rent. What was Zarena's monthly rent before the increase?

whole = part/%
whole = $38 ÷ 9.5%

STEP 3: Estimate the solution.

Estimating before you solve a problem has two benefits:

(1) An estimate will help you determine whether you are using the right formula to solve the problem. Estimates make sense only when you are using the right formula.

(2) A way to check your solution is to compare it to your estimate. If the solution is close in value, it is probably correct.

To estimate the solutions to the example problems, think of the problems this way:

Problem A
The sales tax in Rico's city is 5% of an item's selling price. How much tax did Rico pay on a $43 ladder?

$\frac{1}{20} \times \$40 = \$2.00$

part = 5% × $43

Chapter 3: Working with Percents

Problem B
During a sale, UpTown Clothiers took $18 off the price of an $80 coat. By what percent was the coat marked down?

$$\frac{\$20}{\$80} = \frac{1}{4} = 25\%$$

$$\% = \frac{\$18}{\$80}$$

Problem C
Zarena's 9.5% rent increase added $38 to her monthly rent. What was Zarena's monthly rent before the increase?

$$\$38 \div \frac{1}{10} = \$380$$

whole = $38 ÷ 9.5%

▌ STEP 4: Solve the problem.

Work out the solution using the actual facts from the problem. Notice that the solutions are close in value to the estimates:

Problem A
The sales tax in Rico's city is 5% of an item's selling price. How much tax did Rico pay on a $43 ladder?

part = .05 × $43
part = $2.15
Rico paid $2.15 tax on the ladder.
(The estimate was $2.00.)

Problem B
During a sale, UpTown Clothiers took $18 off the price of an $80 coat. By what percent was the coat marked down?

$$\% = \frac{\$18}{\$80}$$
$$\% = 22.5\%$$

The coat was marked down by 22.5%.
(The estimate was 25%.)

Problem C
Zarena's 9.5% rent increase added $38 to her monthly rent. What was Zarena's monthly rent before the increase?

whole = $38 ÷ .095
whole = $400
Zarena's rent was $400 before the increase.
(The estimate was $380.)

EXERCISE 38

Solve each of these problems. (As a reminder, the steps are given for the first five problems.)

1. Kim bought a new stereo for $510. She paid 6% sales tax. How much was the tax?

    (a) Restate the problem. _____

    (b) Choose the formula to use. _____

(c) Estimate the solution. _____

(d) Solve the problem. _____

2. Joan paid $45 sales tax when she bought a $500 refrigerator. What percent was the sales tax?

   (a) Restate the problem. _____

   (b) Choose the formula to use. _____

   (c) Estimate the solution. _____

   (d) Solve the problem. _____

3. At the Blair Village parking lot, the daily rate recently went up by $1 from $4.50. By what percent did the daily rate rise?

   (a) Restate the problem. _____

   (b) Choose the formula to use. _____

   (c) Estimate the solution. _____

   (d) Solve the problem. _____

4. After Christmas, men's jackets sold for $85. The marked-down price was 68% of the before-Christmas price. What had the jackets cost before Christmas?

   (a) Restate the problem. _____

   (b) Choose the formula to use. _____

   (c) Estimate the solution. _____

   (d) Solve the problem. _____

5. Pinetown recently reduced its city speed limit to 30 miles per hour. The new speed limit is only $66\frac{2}{3}\%$ of the earlier speed limit. What had the earlier speed limit been?

   (a) Restate the problem. _____

   (b) Choose the formula to use. _____

   (c) Estimate the solution. _____

   (d) Solve the problem. _____

6. Celia deposited $5000 in a savings account and left it there for one year. By the end of the year her deposit had earned $287.50 interest. What rate of interest did the bank pay on her deposit?

7. A stationery store bought scenic calendars for $1.20 apiece. It sold the calendars at retail for 375% of the buying price. How much did the store charge for the calendars?

8. Anna bought a new car a year ago. In one year the car's value depreciated (went down) by $16\frac{2}{3}\%$, or $2060. What did she pay for the car when it was new?

9. An 8% sales tax added $24 to the amount Victor paid for a rocking chair. What was the price of the rocking chair before the sales tax was added?

10. A certain clothing store buys shirts for $12. It sells them for $8 more. By what percent does the store mark up the price it pays for the shirts when it sells them?

11. At 8.75%, how much sales tax did Kwon Li pay on a living room set that cost $1450? (Round your answer to the nearest cent.)

12. At a collectors' convention Jerry paid $60 for an old first-edition comic book. The next year he sold it for $37\frac{1}{2}\%$ more. How much had the value of the comic book appreciated (gone up) by the time Jerry sold it?

13. Franco took a one-year loan for $1800 at 11% interest. How much interest did he have to pay on the money he borrowed?

14. Hank's new car cost him $14,280. In addition he paid $856.80 in sales tax. What percent was his sales tax?

15. A lottery winner had to pay $588,000 in taxes on the amount she won. That amounted to 49% of her winnings. How much had she won?

Check your answers on page 239.

Lesson 39

Solving Multistep Percent Problems

Sometimes a percent problem won't give you one of the numbers you need to solve it. You must use the facts you *are* given to find the missing number before you begin to solve the problem.

Finding Parts Larger than the Whole

Many "find the part" problems give you the two numbers you must multiply to solve the problem. For example, if you are asked what the 7% tax on $139 is, you would multiply .07 times $139.

Other "find the part" problems don't give you both the numbers you must use when you multiply. You need to use the facts you are given to find one of the numbers.

The following example shows how to solve such a problem. Notice that the problem doesn't ask for only the tax on the shirt. Rather, it asks the price of the shirt plus the tax.

158   UNIT 2: Percents

Example 1: Abdul bought a $24 shirt. The sales tax was $8\frac{1}{4}\%$. How much did he pay altogether for the shirt?

Step 1

$$100\% + 8\frac{1}{4}\% = 108\frac{1}{4}\%$$

Step 2

part = % × whole
part = 1.0825 × $24
part = $25.98

STEP 1: Abdul paid the price of the shirt (100% × $24) plus the tax on the shirt ($8\frac{1}{4}\%$ × $24). Add the two percents together.

STEP 2: Write the formula for finding the part. Rewrite the formula substituting the sum of the two percents and the price of the shirt. The result is the amount Abdul paid altogether.

In addition to sales tax problems, many other types of problems can be solved by using the method illustrated in Example 1. The following exercise includes a variety of them.

EXERCISE 39a

Solve each problem. Round money to the nearest cent.

1. Tanya bought two sweaters for $90. The tax was 7.5%. How much did she pay in all?

2. One morning in Boulder, Colorado, the temperature was 30°. A warm wind blew in from the mountains, and by 3:00 P.M. the temperature had increased 140% over the morning temperature. What was the temperature at 3:00 P.M.?

3. A manufacturer advertised a new car that would go 20% farther on a gallon of gas than the old model. The old model got 35 miles per the gallon. How many miles per gallon would the new car get?

4. Adrienne bought a new line of dresses to sell in her shop. She paid $62.50 per dress and marked her cost up by 85% to set the retail price. How much did she charge for each dress?

5. Angelo's Service Station purchased several hundred cases of motor oil at $8.95 per case. To set the selling price, Angelo added a 40% markup to each case. How much did Angelo charge for a case of motor oil?

6. Larry buys, repairs, and resells old motorcycles. One motor cycle cost him $550. He sold it at a 45% profit. At what price did he sell the motorcycle?

7. Marcia's supervisor told her that her $18,750 salary would be increased by 5.28%. How much will her salary be after the increase?

8. A year ago Toby bought a gold necklace for $375. Since then, gold jewelry has appreciated 12% in value. How much is his necklace worth now?

9. Last year Marie added $510 to the pension fund she has through her job. In one year the value of the fund appreciated by $8\frac{1}{4}\%$. After a year, what was the value of the money Marie added?

10. One year James sold more merchandise than any other salesperson. His company awarded him a bonus equal to $16\frac{2}{3}\%$ of his $36,000 earnings that year. What was his total income for the year?

Check your answers on page 239.

Finding Parts Smaller than the Whole

Some "find the part" problems tell you the percent by which the whole is decreased. Then they ask how much of the whole is left. Since percents are based on 100, the portion left is always 100% minus the percent of decrease. For example, if a whole (100%) is decreased by 10%, 90% of the whole is left.

The following example shows how to solve such a problem. Notice that the problem gives the percent of a cab's depreciation, or decrease in value. It then asks how much of the original value of the cab is left after it has depreciated.

Example 2: A taxi company buys new cabs at $17,500 each. They depreciate in value by 35% a year. What is the value of a cab after one year?

Step 1	**Step 2**
100% − 35% = 65%	part = % × whole
	part = .65 × $17,500
	part = $11,375

STEP 1: The cab's value depreciated 35% in one year. Subtract 35% from 100% to find what percent of the cab's original value is left.

STEP 2: Write the formula for finding the part. Rewrite the formula substituting the percent of the cab's remaining value and the cab's original value. The result is the amount the cab is worth after one year's depreciation.

In addition to depreciation problems, many other types of problems can be solved by using the method illustrated in Example 2. The following exercise includes a variety of them.

EXERCISE 39b

Solve each problem. Round money answers to the nearest cent.

1. Smith Construction Company bought a bulldozer for $150,000. A year later it had depreciated 15% in value. What was the bulldozer's value after a year?

2. The Miller family purchased a co-op apartment for $111,000. Soon after that, real estate values in their city fell. The Millers' co-op's value decreased by $33\frac{1}{3}$%. What was the co-op worth after its value fell?

3. Sylvia wants to buy a word processor. The usual price is $570, but this month they are on sale for 25% off. If Sylvia buys a word processor this month, how much will she pay?

4. Arrow Autos advertised last year's $12,550 models for 20% off during a clearance sale. What was the sale price for one of last year's cars?

5. Jason bought a sofa in the store that had this sign in the window. The sofa's original price had been $776. What was its sale price?

    FIRE SALE
    All Merchandise
    40% Off

6. OnTime Airlines offered a $37\frac{1}{2}$% discount on its $1090 flights between Miami and Los Angeles. How much was the discounted fare?

7. In February, $15 appointment books went on sale for 50% off. What was the sale price of the appointment books?

8. To save their employer from bankruptcy, the employees agreed to a 12% pay cut. Juan's salary had been $24,500. What was his salary after the cut?

9. When the Middletown Savings and Loan collapsed, the depositors took a 65% loss on their accounts. Blanca had $2900 in the bank. How much did she receive after the collapse?

10. Mary invested in some bad stock. She paid $34.57 per share for it, but by the end of one month its value had fallen by 11%. How much was Mary's stock then worth per share?

Check your answers on page 240.

Finding Percents of Change

Often when a problem asks you to find a percent, it gives you the numbers you must divide to solve them. For example, a problem may ask by what percent the value of an $85,000 house rose if its worth went up by $12,750. To find the answer, you would divide $12,750 by $85,000.

Chapter 3: Working with Percents   161

Other "find the percent" problems don't give you both the numbers you must use when you divide. You need to use the facts you are given to find one of the numbers.

The following example shows how to solve such a problem. Notice that the problem doesn't ask what percent of the original value the new value of the plate is. Nor does it ask the opposite. Rather, it asks what percent of the plate's original value the amount of appreciation was.

Example 3: Sam bought a commemorative plate for $20. Three years later he sold it for $24. By what percent did the value of the plate appreciate in three years?

Step 1

$24 − $20 = $4

Step 2

% = part ÷ whole
% = $4 ÷ $20
% = .2

.2 = 20%

STEP 1: To find how much the plate's value appreciated, subtract its original value from its new value. The plate appreciated $4 in value.

STEP 2: Write the formula for finding the percent. Rewrite the formula substituting the amount of appreciation and the plate's original value. The result is the percent by which the value of the plate appreciated.

You can use the method shown in Example 3 to solve any problem that asks for the percent of change when that change is an *increase*.

To find the percent of change when the change is a *decrease*, the method for solving a problem is similar. Example 4 illustrates the method.

Example 4: Jamal usually charges $16 for electric screwdrivers. This week he put the screwdrivers on sale for $11.20. By what percent did Jamal discount the price of the screwdrivers?

Step 1

$16.00 − $11.20 = $4.80

Step 2

% = part ÷ whole
% = $4.80 ÷ $16
% = .3

.3 = 30%

STEP 1: To find how much Jamal discounted the price of the screwdrivers, subtract the sale price from the original price. The discount amounted to $4.80.

STEP 2: Write the formula for finding the percent. Rewrite the formula substituting the amount of the discount and the screwdrivers' original price. The result is the percent of the discount.

All the problems in the following exercise ask for the percent of change. Read them carefully because some ask about increases and some ask about decreases.

EXERCISE 39c

Find the percent of change in each problem.

1. A bookstore bought a supply of almanacs for $1.25 each and sold them for $5.95. By what percent did the bookstore mark up the almanacs?

2. Peggy bought an antique table for $240 at an auction. Later she sold the table in her shop for $360. By what percent had Peggy marked the table up?

3. At another auction Peggy bought a painting for $100. A few months later she sold it for $80. By what percent did Peggy reduce the cost of the painting to sell it?

4. Audrey's father left her an acre of land valued at $7200. She sold it recently for $7920. What percent profit did Audrey realize when she sold the land?

5. On Saturday Allen picked tomatoes at his aunt's farm and paid her $2.20 for each basket he picked. On Sunday, he sold some of the tomatoes at his vegetable stand for $5.50 a basket. By what percent did Allen mark up the tomatoes?

6. On Monday Allen sold the remainder of the tomatoes for $1.10 a basket to get rid of them. By what percent did Allen mark down his cost for the tomatoes below his cost? (Remember: Allen had paid $2.20 a basket for the tomatoes.)

7. Suppose you bought a rare coin some years ago for $250. You heard that recently a coin just like yours was sold at a coin show for $750. If your coin is also worth $750 now, by what percent has its value increased?

8. Peter bought a ton of scallops for $3.36 a pound at a wholesale market. He sold them to a supermarket chain for $3.92 a pound. By what percent did Peter mark up his cost for the scallops?

9. A sporting goods store had this sign in its window. By what percent had the store discounted the exercise shoes?

    EXERCISE SHOES
    Were $60
    Now $39

10. When Joaquin started working as an electric cable splicer, his wage was $11.60 per hour. Thanks to a raise, he now makes $12.18 per hour. By what percent has his wage gone up?

Check your answers on page 241.

PERCENTS REVIEW

Part A. Complete the statements about the following figures.

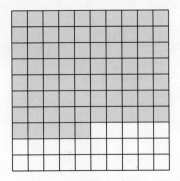

Figure 1

1. _____ % of Figure 1 is shaded.

2. _____ % of Figure 1 is not shaded.

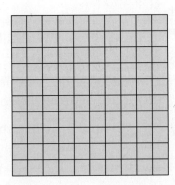 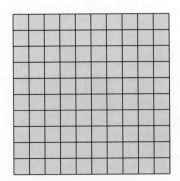

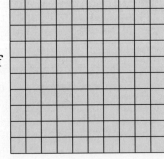 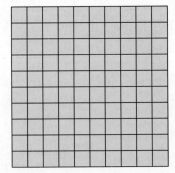

3. In Figure 2, _____ % of one large square is shaded.

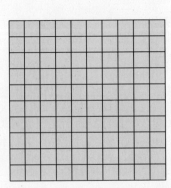 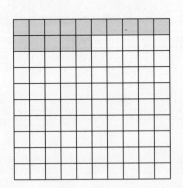

Figure 2

164   UNIT 2: Percents

4. _____ % of Figure 3 is shaded.

5. _____ % of Figure 3 is not shaded.

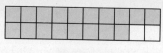

Figure 3

Part B. Rewrite each number as a percent.

1. .03   2. .0825   3. $\frac{1}{4}$ 4. $\frac{2}{10}$ 5. $\frac{6}{16}$

Part C. Fill in the blanks.

	Percent	Fraction	Decimal
1.	50%	_____	_____
2.	_____	_____	.75
3.	_____	$\frac{1}{3}$	_____
4.	87.5%	_____	_____
5.	_____	$\frac{5}{6}$	_____
6.	_____	_____	.2
7.	_____	_____	$.16\frac{2}{3}$
8.	62.5%	_____	_____
9.	_____	$\frac{1}{10}$	_____
10.	_____	$\frac{1}{8}$	_____

Part D. Solve each problem.

1. What is 60% of 80?
2. What percent of 36 is 9?
3. What is 50% of 128?
4. What percent of $35 is $17.50?
5. What is 16% of 150?
6. What is $66\frac{2}{3}$% of 96?
7. 9 is $16\frac{2}{3}$% of what number?
8. What is .5% of 400?
9. 250% of what number is 25?
10. What percent of 1,000,000 is 100?

Percents Review   165

Part E. Solve each of the following word problems.

1. Estimate the answer to the following: Last year the home team played well, and the average attendance at games was 5983. This year the home team is not playing well, and the average attendance at games is only 2000. This year's average attendance is *about* what percent of last year's?

2. Salesclerks at Clark's Carpet City get a 9% commission on sales. Melanie sold a $432 carpet this morning. What is her commission on that sale?

3. Boys' $125 suits went on sale for $25 off. By what percent were the suits marked down?

4. Falcon bought a new car last year. Its value has depreciated so that the car is now worth only $12,000. That is 75% of the car's original value. What did Falcon pay for the car?

5. Because of a lower speed limit, Lumberton County reported a 36% decrease in highway fatalities this year compared to last year. Last year there were 150 highway deaths. How many were there this year?

6. Jeremy bought a $15,000 car and paid an 8.75% sales tax. How much did he pay altogether for the car?

7. Li buys dish-towel calendars for $1.50 apiece. He sells them to commuters in subway cars for $5.00. By what percent does Li mark up his cost for the calendars?

8. A discount store ran this ad in a newspaper. By what percent had the store discounted the paint?

LATEX PAINT
Was $15 a gallon
Now $12

9. George got 15 of the 18 problems in an exercise correct. What percent of the problems did he have right?

10. Felix got a 5.95% pay increase. His old salary was $13,750. What will his new salary be? (Round your answer to the nearest cent.)

Check your answers on page 241.

166   UNIT 2: Percents

GED PRACTICE 2

This section will give you practice in answering questions like those on the GED. The 15 questions in this Practice are multiple-choice like the ones on the GED. As you do this Practice, use the skills you've studied in this unit and follow these steps with each problem:

- Restate the problem.
- Choose the formula to use.
- Estimate the solution.
- Solve the problem.

Directions: Choose the one best answer to each item.

1. Constantine purchased an electronic typewriter for $180. The sales tax was $7\frac{1}{2}$%. How much did he pay?
   (1) $ .04
   (2) $ 13.50
   (3) $ 24.00
   (4) $ 193.50
   (5) $1350.00

2. Sterling Sounds sold $3500 worth of CDs in one day. The store collected $227.50 in sales tax on those sales. What was the percent of the sales tax?
   (1)   6.5%
   (2) 14.38%
   (3) 15.38%
   (4) 93.5%
   (5) 106.5%

3. Tina borrowed some money for one year at 8% interest. When she repaid the loan, the interest due was $128. How much money had Tina borrowed?
   (1) $ 10.24
   (2) $ 16.00
   (3) $1024.00
   (4) $1600.00
   (5) $1728.00

Item 4 refers to the following diagram.

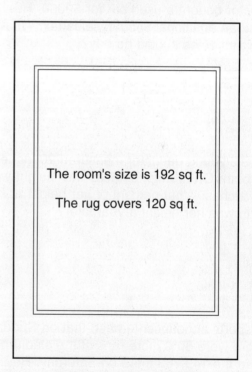

The room's size is 192 sq ft.

The rug covers 120 sq ft.

Thelma's Living Room

4. What percent of Thelma's living room floor is covered by her rug?
   (1)   1.6%
   (2)   2.3%
   (3) 37.5%
   (4)   60%
   (5) 62.5%

5. In a class called "Asian Cooking," 20% of the students have never cooked before. How many of the 45 students in the class are learning to cook for the first time?
   (1) 225
   (2) 54
   (3) 45
   (4) 36
   (5) 9

6. Gregory received a check for $1000.00, which was the $16\frac{2}{3}$% dividend earned by some money he had invested. How much money had Gregory invested?
   (1) $ 166.67
   (2) $1166.67
   (3) $6000.00
   (4) $6161.43
   (5) $7000.00

7. Hector bought a used car for $9000. He paid an additional $450 in sales tax. What percent sales tax did he pay?
   (1) .05%
   (2) 5%
   (3) 19%
   (4) 20%
   (5) 21%

8. Luis's young daughter measured 20 inches at birth. On her first birthday, her height was 28 inches. What percent of her height at birth was her height a year later?
   (1) 28.6%
   (2) 40%
   (3) 71.4%
   (4) 140%
   (5) 180%

9. A sports announcer learned that on Sunday there were 50% more fans at the stadium than the 149,763 there on Saturday. He estimated the size of Sunday's crowd and later reported that it numbered about
   (1) 300,000
   (2) 225,000
   (3) 150,000
   (4) 100,000
   (5) 75,000

10. A newsstand buys newspapers for $.20 apiece and sells them with a 150% markup. How much does each newspaper sell for?
    (1) $.10
    (2) $.20
    (3) $.30
    (4) $.40
    (5) $.50

11. Martha's base wage is $8 per hour. When she works the night shift, she makes $8.50 per hour. By what percent does Martha's base wage increase when she works the night shift?
    (1) 5.88%
    (2) 6.25%
    (3) 12.5%
    (4) 94.12%
    (5) 106.25%

12. A news report said that the price of milk had risen $.10 a quart—up from $.79. The report estimated the percent of the increase at
    (1) .125%
    (2) 8%
    (3) 10%
    (4) 12.5%
    (5) 79%

13. A certain black-and-white television regularly sells for $280. During a sale, the television's price was discounted 20%. What was the television's sale price?
    (1) $ 5
    (2) $ 14
    (3) $ 56
    (4) $224
    (5) $336

14. Mike's diet helped him to go from 250 pounds to 230 pounds in one month. By what percent did his weight decrease during that month?
    (1) 92%
    (2) 12.5%
    (3) 11.5%
    (4) 8.7%
    (5) 8%

15. Anna bought a plot of land a year ago for $4800. Since then, the plot's value has appreciated 4%. How much is the plot worth now?
    (1) $ 192
    (2) $1200
    (3) $4608
    (4) $4992
    (5) $6000

Check your answers on page 242.

GED PRACTICE 2 SKILLS CHART

To review the mathematics skills covered by the items in GED Practice 2, study the following lessons in Unit 2.

Unit 2	Percents	Item Number
Lesson 34	Finding the Part	1, 5, (9, 10, 13, 15)
Lesson 35	Finding the Percent	2, 4, 7, 8, (11, 12, 14)
Lesson 36	Finding the Whole	3, 6
Lesson 37	Estimating Solutions to Percent Problems	9, 12
Lesson 38	Solving One-Step Percent Word Problems	(1–8, 12)
Lesson 39	Solving Multistep Percent Problems	(9), 10, 11, 13, 14, 15

UNIT 3

Tables and Graphs

Chapter 1 of this unit introduces tables and graphs. In Chapter 2 you will learn to solve one-step and multistep problems based on tables and graphs.

Unit 3 Overview

Chapter 1 Understanding Graphic Information
Chapter 2 Working with Graphic Information

GED Practice 3

Chapter 1
UNDERSTANDING GRAPHIC INFORMATION

The four lessons in this chapter introduce graphic displays: tables, bar graphs, pictographs, line graphs, circle graphs, and divided bar graphs. You will learn what each type of graphic display is used for and how to read and get information from each. These lessons cover some of the skills you will use in Chapter 2 of this unit.

Lesson 40

Graphic Displays of Data

When numerical information, or **data**, is given in a paragraph, it is not easy to find a particular piece of information or to compare it to other data. For example, if you want to know which has more calories, lima beans or kidney beans, it would take some time to find the answer by reading this paragraph:

> Most grains, vegetables, and beans have few calories. A half-cup serving of green peas has only 60 calories. Kidney beans have 110 calories per half cup. A half cup of corn has 85 calories. In half-cup servings of lima beans and oatmeal there are 65 and 80 calories, respectively.

Certain graphic displays make it easier to find and compare data. Three types of displays are common: tables, bar graphs, and pictographs. Each type has its own special benefit. As you work through this lesson, you will see the calorie data shown in each of these three types of graphic displays.

Tables

Tables are used to display data when it is important to show exact amounts. In a table, numbers are arranged in **columns** and **rows**. A table's **title, subtitle, column headings,** and **row headings** make the meaning of the numbers clear. Sometimes a table has an explanatory **footnote**. Often there is a **source** line to tell where the data in the table came from.

The following table displays the data from the calorie paragraph above. The parts of the table are labeled.

Title → **CALORIE CONTENT OF CERTAIN FOODS**
Subtitle → Serving Size: ½ Cup

Column → Food   Calories*
headings

Row headings:
Corn   85
Green peas   60
Kidney beans   110
Lima beans   65
Oatmeal   80

Footnote → *Rounded to the nearest 5.
Source → *Source:* U.S. Department of Health and Human Services.

The title shows what the table displays: the number of calories in certain foods.

The subtitle says that the numbers in the table report the amount of calories in one-half cup of each food.

The column headings, *Food* and *Calories*, tell what each column lists.

The row headings are the names of five foods. In this table the row headings are in alphabetical order to make it easy to find a particular food.

The footnote explains that the calorie amounts are rounded to the nearest 5. For example, corn could have anywhere from 83 to 87 calories in a half cup, but the table reports 85.

The source line tells what agency provided the data for the table.

Most tables report more information than the calories table. They usually have more than two columns. The table in Example 1 has five columns.

To find a piece of information in a table, find the number where the appropriate row and column meet.

Example 1: According to the following table, what was the record for the number of passes completed by an NFL player in 1983?

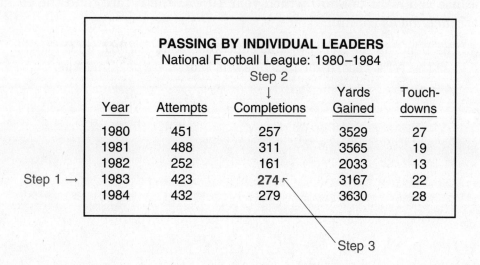

PASSING BY INDIVIDUAL LEADERS
National Football League: 1980–1984

Step 2 ↓

Year	Attempts	Completions	Yards Gained	Touch-downs
1980	451	257	3529	27
1981	488	311	3565	19
1982	252	161	2033	13
1983	423	274	3167	22
1984	432	279	3630	28

Step 1 → 1983

Step 3

Chapter 1: Understanding Graphic Information

STEP 1: Because the question asks about 1983, find the row labeled *1983*.

STEP 2: Because the question asks about the number of passes completed, find the column labeled *Completions*.

STEP 3: The number where the row and column meet is *274*. Therefore, the record for the number of passes completed by an NFL player in 1983 was 274.

If you know a piece of information and want to find its category, locate the information in the table and then look for the appropriate row or column heading.

Example 2: You know that one year the NFL record for touchdowns scored by passing was 19, but you don't remember the year. Using the following table, find the year.

PASSING BY INDIVIDUAL LEADERS
National Football League: 1980–1984

Year	Attempts	Completions	Yards Gained	Touch-downs
1980	451	257	3529	27
1981	488	311	3565	19
1982	252	161	2033	13
1983	423	274	3167	22
1984	432	279	3630	28

Step 1 → Touch-downs column
Step 2 ← 19
Step 3 → 1981

STEP 1: Because you want to know about a touchdown record, find the column labeled *Touchdowns*.

STEP 2: Because you want to know which year 19 was the record, find the *19* in the *Touchdowns* column.

STEP 3: Find the heading for the row that contains *19*. It is *1981,* so that was the year the NFL record for touchdowns scored by passing was 19.

EXERCISE 40a

Answer the questions about the following table.

PRINCIPAL CAUSES OF ACCIDENTAL DEATHS, U.S.A.

Year	Motor Vehicles	Falls	Fires/ Burns	Drowning	Guns	Choking	Poisons
1987	52,600	12,300	5,500	7,000	1,800	3,249	4,300
1988	46,000	11,600	5,000	6,200	1,900	3,200	4,400
1989	46,200	11,800	5,100	5,600	1,700	3,100	4,300
1990	47,900	11,000	4,800	5,600	1,800	3,600	4,900

Note: Number of deaths is rounded to the nearest hundred.
Source: National Safety Council.

1. The table shows information about what?
2. Where did the information in the table come from?
3. How many fatal gun accidents were there in 1988?
4. For each year shown in the table, which kind of accident caused the most deaths?
5. In what year were more than 12,000 fatalities caused by falls?
6. Poisons caused 4,300 deaths in which two years?
7. For every year shown in the table, which kind of accident caused the fewest deaths?
8. Fewer than 5,000 people were killed by burns in which year?
9. In 1990, how many people died by choking?
10. What single type of accident resulted in 5,600 deaths in 1989?

Check your answers on page 243.

Chapter 1: Understanding Graphic Information

Bar Graphs

Bar graphs are used to display data when it is more important to show comparisons than exact amounts. The **bars** on a **vertical-bar graph** extend upward from its **horizontal axis**. The **bars** on a **horizontal-bar graph** extend toward the right from its **vertical axis**. The **title**, the **subtitle**, the **scale** and the **labels** on the two **axes** and on the **bars** make the meaning of the graph clear. Like tables, bar graphs often have an explanatory **footnote** and a **source** line.

The following bar graphs display the data from the calorie paragraph on page 172. On the left is a **vertical-bar graph**; on the right, a **horizontal-bar graph**. Both graphs show the same information. Their parts are labeled.

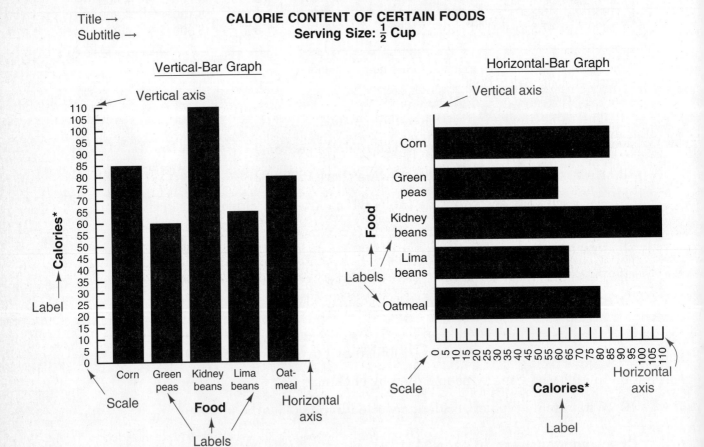

Footnote → *Rounded to the nearest 5.
Source → *Source:* U.S. Department of Health and Human Services.

As in tables the title and subtitle tell what the bar graph shows. A footnote explains that the graph reports rounded amounts and a source line tells where the data in the graph came from.

On the vertical-bar graph the label on the vertical axis shows that the numbers on its scale refer to calories. On the horizontal-bar graph calories are shown on the horizontal axis. On both graphs the scales are labeled in increments of 5.

176 UNIT 3: Tables and Graphs

On the vertical-bar graph, the label on the horizontal axis shows that the bars represent types of food. On the horizontal-bar graph the food label is on the vertical axis. On both graphs each bar has a label to tell which type of food it represents.

Bar graphs make it easy to compare data. If you want to find the largest amount shown, look for the longest bar. To find the smallest amount look for the shortest bar. Look at the lengths of the other bars to see how the amounts they stand for compare to the largest and smallest amounts.

Example 3: According to the following graph, falls account for the highest number of non-vehicle accidental deaths in 1990. What kind of accident resulted in the second highest number of deaths?

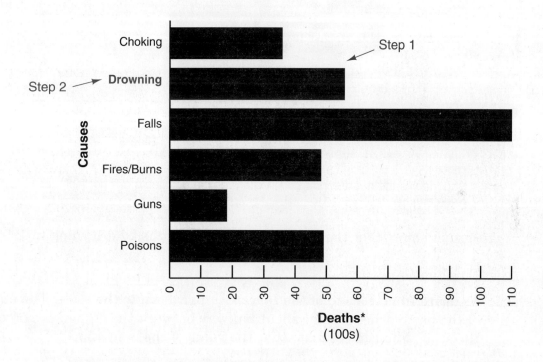

*Rounded to the nearest hundred. Motor vehicle accidents are not included.
Source: National Safety Council.

STEP 1: Because the question asks about the *second* highest number of deaths, find the bar that is next in length to the bar labeled *Falls*.

STEP 2: Look at the label on that bar. It tells you that drowning accounted for the second highest number of deaths in 1990.

The scale on a bar graph makes it possible to read data from the graph. To find a piece of information, draw an imaginary line to the scale from the top or end of the bar whose value you want to know. If the bar is not even with one of the points on the scale, estimate. (Because scales do not show every number in a range, bar graphs are usually not as accurate as tables, which can display any number.)

Chapter 1: Understanding Graphic Information   177

Example 4: About how many drowning deaths were there in 1990?

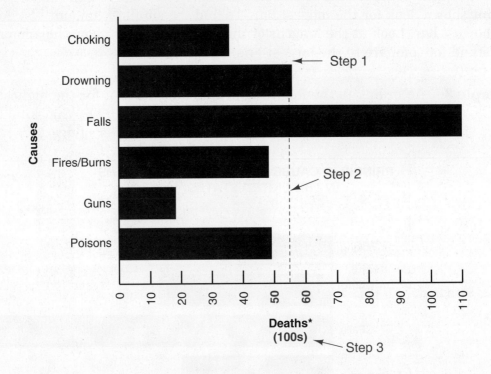

*Rounded to the nearest hundred. Motor vehicle accidents are not included.
Source: National Safety Council.

The graph shows that there were about 5500 deaths by drowning in 1990.

STEP 1: Find the *Drowning* bar.

STEP 2: Measure from the end of the *Drowning* bar down to the scale. The end of the bar reaches a point about halfway between the *50* and the *60* on the scale. You can estimate that the value of the bar is 55.

STEP 3: Notice that the label on the horizontal scale includes *(100s)*. That means that each number on the scale stands for itself times 100. Therefore, there were about 5500 deaths by drowning in 1990 because 55 × 100 = 5500. (5500 is a close estimate; the actual number of deaths by drowning was 5600.)

EXERCISE 40b

Answer the questions about the following bar graph.

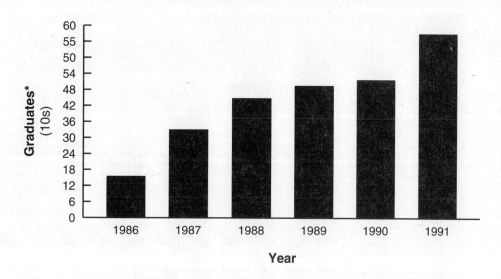

*Rounded to the nearest 10.
Source: *The County GED Program: Success Rate*, a report by the Middle County School Board.

1. What information does the graph report?
2. Where did the information in the graph come from?
3. In which year were there the fewest GED graduates?
4. In which year were there the most GED graduates?
5. The number *18* on the scale stands for how many graduates?
6. About how many graduates were there in 1987?
7. About how many graduates were there in 1990?
8. In which year were there about 450 graduates?
9. Between which two years was the increase in the number of graduates the greatest?
10. Between which two years was the increase in the number of graduates the smallest?

Check your answers on page 243.

Pictographs

Of the three kinds of graphs used to display data, pictographs provide the strongest visual image. **Symbols**, or pictures, represent the data the graph shows. Each symbol on a pictograph stands for a particular amount of a certain item. The **title** and the **subtitle** tell what the graph shows. A **key** explains the value of each symbol, and **labels** next to each **row** of symbols tell what item the symbols in that row stand for. Like tables and bar graphs, pictographs often have an explanatory **footnote** and a **source** line.

The following pictograph displays the data from the calorie paragraph on page 172. The parts of the graph are labeled.

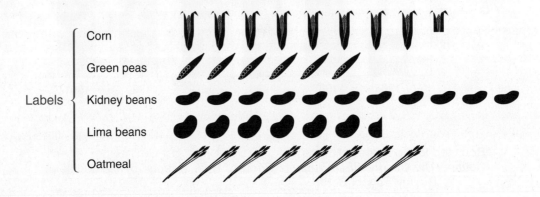

Title → **CALORIE CONTENT OF CERTAIN FOODS**
Subtitle → **Serving Size: $\frac{1}{2}$ Cup**

Labels: Corn, Green peas, Kidney beans, Lima beans, Oatmeal

Key → **Key:** Each symbol stands for 10 calories.
Footnote → *Rounded to the nearest 5.
Source → *Source:* U.S. Department of Health and Human Services.

As in tables and graphs, the title and subtitle tell what the pictograph shows. A footnote explains that the graph reports rounded amounts, and a source line tells where the data in the graph came from. The label before each row of symbols tells what the symbols stand for.

The key says that each symbol on the graph stands for 10 calories. Therefore, wherever there is only half a symbol on the graph, it stands for 5 calories.

With pictographs it is easy to compare data by counting the symbols in each row or looking at the lengths of the rows. The key makes it possible to read data from a pictograph. (Pictographs are usually drawn with only whole symbols and half symbols, so they show amounts with less accuracy than bar graphs and far less accuracy than tables.)

Example 5: According to the following pictograph, the number of GED graduates was about equal in which two years? How many people earned GEDs in each of those years?

180 UNIT 3: Tables and Graphs

MIDDLE COUNTY SCHOOL DISTRICT
Number of GED Graduates: First 6 Years of Programs

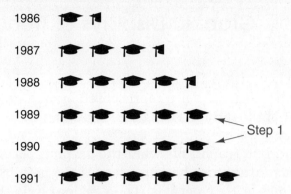

Key: Each symbol stands for 10 graduates. ← Step 2
Note: Number of graduates is rounded to the nearest 10.
Source: The County GED Program: Success Rate, a report by the Middle County School Board.

STEP 1: The *1989* and *1990* rows have the same number of symbols. Therefore the number of graduates must have been about equal in those two years.

STEP 2: The key says that each symbol stands for 10 graduates. Since there are 5 symbols in both the *1989* and *1990* rows, there were about 50 graduates in each of those years. (Because pictographs do not report accurate figures, 50 is only an estimate for 1990. The actual number of 1989 graduates *was* 50, but the 1990 number was 52.)

EXERCISE 40c

Answer the questions about the hog-production pictograph.

1. What information does the graph report?
2. Where did the information in the graph come from?
3. What do the labels on the rows in the graph stand for?
4. Which county had the lowest hog production?
5. Which county had the highest hog production?
6. Which two counties produced about the same number of hogs?
7. Each symbol in the graph stands for how many hogs?
8. About how many hogs were produced in Davis County?
9. About how many hogs were produced in Cook County?
10. In which county were about 55,000 hogs produced?

Check your answers on page 243.

Lesson 41: Graphic Displays of Trends

Line graphs display data. They usually show how something has changed over time. The data on line graphs are represented by **points on the line**. The **line** shows the **trend** of the data: a rising line shows an increase; a falling line shows a decrease.

The **title** of a line graph tells what the graph reports. Both axes have **labels**. The **horizontal axis** that runs across the bottom of a line graph usually shows times or time periods. The **vertical axis** that runs up the left side of the graph gives amounts. The **scale** on the vertical axis shows the value of the points on the line. A line graph may have a **subtitle**, an explanatory **footnote**, and a **source** line.

The following line graph displays some of the data from the table on page 175. Its parts are labeled.

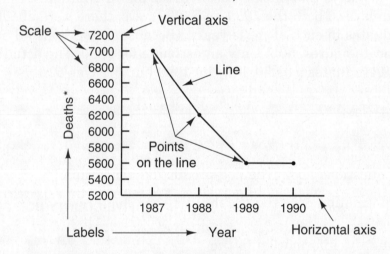

Title → **DEATHS CAUSED BY ACCIDENTAL DROWNING IN THE UNITED STATES, 1987–1990**

Footnote → *Rounded to the nearest hundred.
Source → *Source:* National Safety Council.

The title tells that the line graph shows how many deaths were caused by accidental drowning in the United States from 1987 to 1990. A footnote explains that the graph reports deaths in rounded amounts. A source line tells where the data in the graph came from.

The label on the horizontal axis shows that the numbers along it stand for years.

According to the label on the vertical axis, the numbers on its scale stand for deaths. Notice that the lowest number on the scale is 5200 and that the scale uses increments of 200.

From 1987 to 1989, the line on the graph falls. That shows a decreasing trend: in each successive year there were fewer deaths by accidental drowning. From 1989 to 1990 the line neither rises nor falls. That shows a level trend: the numbers of accidental drowning deaths in 1989 and 1990 were the same.

Line Graphs: Finding Information

To find a piece of information on a line graph, locate on the line the point you want to know about. Trace across to the vertical scale from that point. If the point is not even with one of the numbers on the vertical scale, estimate.

Example 1: According to the following graph, what was the temperature at noon on Monday?

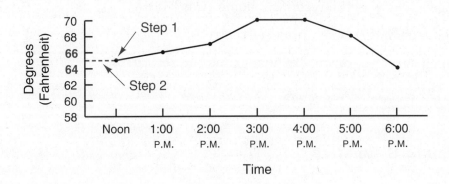

STEP 1: Find the point on the line that stands for noon.

STEP 2: Measure from the *Noon* point over to the scale. The point reaches the scale about halfway between the *64* and the *66*. You can estimate that the temperature at noon was about 65°F.

EXERCISE 41a

Answer the questions about the following line graph.

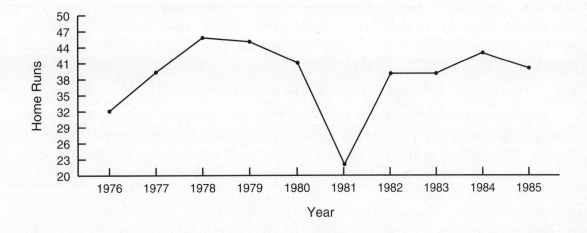

Chapter 1: Understanding Graphic Information

1. What information does the graph report?
2. What does each number on the vertical scale stand for?
3. What does each number on the horizontal scale stand for?
4. In which year was the record number of home runs the greatest?
5. From 1976 to 1978 does the graph show an upward, a level, or a downward trend?
6. What kind of trend does the graph show from 1978 to 1981?
7. What increments does the vertical scale use?
8. In which year was the record 22 home runs?
9. In 1980 how many home runs set the record?
10. In which three years was 39 home runs the record?

Check your answers on page 243.

Line Graphs: Estimating

The scale on a line graph makes it possible to estimate data from any point on the graph's line, even if that point has no label on the horizontal axis.

Example 2: Estimate what the temperature was at 4:30 P.M. on Monday.

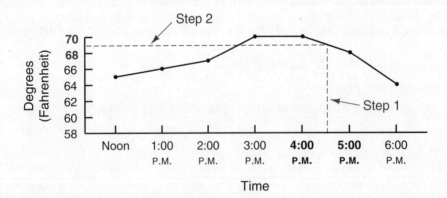

STEP 1: Estimate where 4:30 P.M. would be on the horizontal scale. Measure from there up to the line on the graph.

STEP 2: Measure from the 4:30 P.M. point on the line over to the vertical scale. The 4:30 point reaches the scale about halfway between the *68* and the *70*. You can estimate that the temperature at 4:30 P.M. was about 69°F.

EXERCISE 41b

Answer the questions about the following line graph.

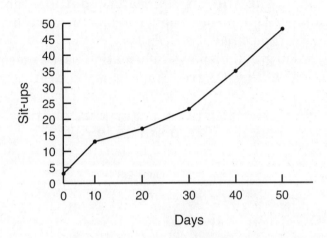

1. Before Laura began to exercise (*0* on the horizontal scale), how many sit-ups could she do?

2. What kind of trend does the graph show?

3. How many sit-ups did Laura do on the thirtieth day of her exercise program?

4. Estimate how many sit-ups Laura probably did on the fifth day of her exercise program.

5. By which day of her exercise program was Laura probably able to do 29 sit-ups?

Check your answers on page 243.

Graphic Displays of the Parts of Wholes

Some graphic displays are used to show how a whole is divided into its parts and how the parts compare in size. Such displays do not usually show data. Rather, they show the sizes of the parts of the whole amount.

Two types of graphic displays are common for showing the portions of a whole, **circle graphs** and **divided-bar graphs**.

Chapter 1: Understanding Graphic Information   185

Circle Graphs

A circle graph is made of a circle divided into sections. (Circle graphs are often called pie charts because they look like pies cut into pieces.) Each section of a circle graph stands for one part of a whole. By comparing the sizes of the sections of a circle graph, you can compare the sizes of the amounts they stand for.

As with other graphs, the **title** of a circle graph tells what it reports. A circle graph may have a **subtitle**, an explanatory **footnote**, and a **source** line. Each section of a circle graph has a **label**. The label usually includes a **percent** to show what portion of the circle it stands for.

The following circle graph displays information based on some of the data in the table on page 175. Its parts are labeled.

Title → **PRINCIPAL CAUSES OF ACCIDENTAL DEATHS IN THE UNITED STATES IN 1990**

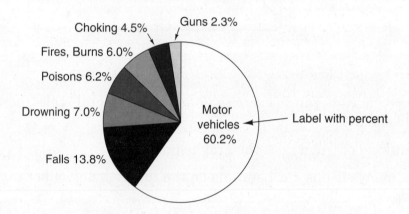

Footnote → *Note:* Number of deaths is rounded to the nearest .1%.
Total deaths reported: 79,600.
Source → *Source:* National Safety Council.

The title tells that the circle graph shows the causes of accidental deaths in the United States in 1990. A footnote explains that the figures in the graph have been rounded to the nearest tenth of a percent. It also explains that the graph's information is based on 79,600 reported accidental deaths. A source line tells where the data in the graph came from.

The whole circle represents all 79,600 accidental deaths in 1990. Each section of the circle stands for one cause of accidental death, as the labels show. The percent of each cause is part of each section label. The size of each section represents the percent it stands for. For example, the section of the graph that stands for deaths caused by motor vehicle accidents makes up 60.2% of the circle.

Circle graphs make it easy to compare the size of the parts of a whole and to find what percent each part is of the whole. If you want to know which part of a whole is the largest, look for the largest section of the circle. To find the smallest part, look for the smallest section. Look at the other sections to see how their sizes compare to the largest and smallest. To find what percent a section stands for, look at its label.

Example 1: According to the following graph, small appliances consume the smallest portion of the electricity an average family uses. What is the second smallest consumer? What percent does it consume?

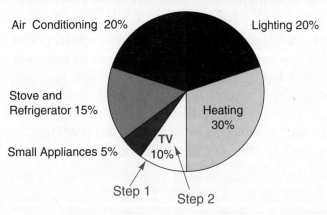

AVERAGE ANNUAL ELECTRICITY CONSUMPTION
Family of Four, 1980–1989

Source: State Commission on Electricity Consumption.

STEP 1: Because the question asks about the *second* smallest consumer of electricity, find the section that is next larger in size to the one labeled *Small Appliances*.

STEP 2: Look at that section's label. It tells you that the TV is the second smallest consumer. It consumes 10% of the electricity used.

EXERCISE 42a

Answer the questions about the energy-sources circle graph.

1. What does the graph show?
2. Where did the information in the graph come from?
3. What does the whole circle stand for?
4. What is the largest source of energy used in the United States?
5. What percent of U.S. energy comes from burning coal?
6. Which source supplies 6.9% of the energy used in the United States?
7. Does the graph show what percent of the energy used in the United States is solar?

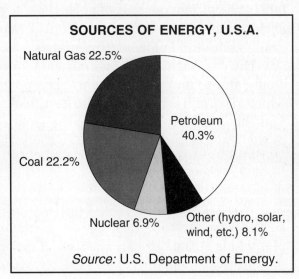

SOURCES OF ENERGY, U.S.A.

Natural Gas 22.5%
Petroleum 40.3%
Coal 22.2%
Nuclear 6.9%
Other (hydro, solar, wind, etc.) 8.1%

Source: U.S. Department of Energy.

Check your answers on page 244.

Divided-Bar Graphs

A divided-bar graph is similar to a circle graph. It is made of a bar—either vertical or horizontal—divided into sections. Each section of a divided-bar graph stands for one part of a whole. By comparing the sizes of the sections of a divided-bar graph, you can compare the sizes of the amounts they stand for.

As with other graphs, the **title** of a divided-bar graph tells what it reports. A divided-bar graph may have a **subtitle**, an explanatory **footnote**, and a **source** line. Each section of a divided-bar graph has a **label**. The label usually includes a **percent** to show what portion of the bar it stands for.

The following divided-bar graph displays the same information as the circle graph at the top of page 187. Its parts are labeled.

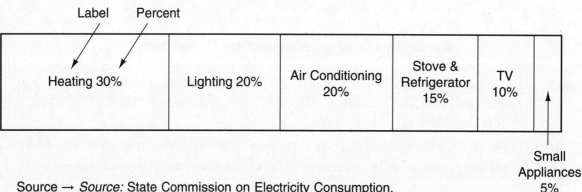

The title and subtitle tell that the divided-bar graph shows how a family of four uses electricity during a year based on a 10-year study. A source line tells where the data in the graph came from.

The whole bar represents all the electricity a family consumes in an average year. Each section of the bar stands for one type of consumption, as the labels show. The percent of each type is given in the section labels. The size of each section represents the percent it stands for. For example, the section of the graph that stands for heating makes up 30% of the whole bar.

Divided-bar graphs make it easy to compare the sizes of the parts of a whole and to find what percent each part is of the whole. If you want to know which part of a whole is the largest, look for the largest section of the bar. To find the smallest part, look for the smallest section. Look at the other sections to see how their sizes compare to the largest and smallest. To find what percent a section stands for, look at its label.

188   UNIT 3: Tables and Graphs

Example 2: According to the following graph, which group of SSI recipients is the smallest? That group makes up what percent of all SSI recipients?

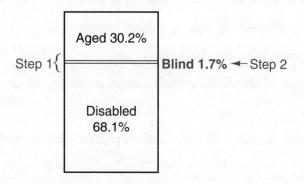

RECIPIENTS OF SUPPLEMENTAL SECURITY INCOME (SSI)
December 1990

Source: Social Security Administration.

STEP 1: Find the smallest section of the graph.

STEP 2: Look at that section's label. It tells you that the blind are the smallest group of SSI recipients. They make up 1.7% of all SSI recipients.

EXERCISE 42b

Answer the questions about the following divided-bar graph.

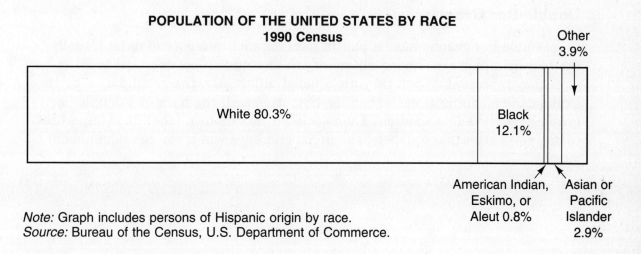

POPULATION OF THE UNITED STATES BY RACE
1990 Census

Note: Graph includes persons of Hispanic origin by race.
Source: Bureau of the Census, U.S. Department of Commerce.

Chapter 1: Understanding Graphic Information   189

1. What does the graph show?
2. In what year was the information for the graph gathered?
3. What organization supplied the information for the graph?
4. What does the whole bar stand for?
5. What percent does the whole bar represent?
6. Which group makes up the smallest segment of the U.S. population?
7. Which group makes up the largest segment of the U.S. population?
8. What percent of the U.S. population is made up of persons who are black?
9. Which group makes up 2.9% of the U.S. population?
10. What percent of the U.S. population is categorized as *Other*?

Check your answers on page 244.

Double Graphs

Double graphs show two types of data for each of several categories. They make it possible to compare a lot of data in a variety of ways. Two kinds of double graphs are common, **double-bar graphs** and **double-line graphs.**

Double-Bar Graphs

Double-bar graphs have a pair of bars for each category of data. Usually the bars in a pair are shaded differently to distinguish between the types of data they represent. When they are shaded differently, the graph has a key to explain the meaning of each shading. Sometimes all the bars in a double-bar graph have the same shading. Then each bar has its own label. It is important to pay close attention to labels, shadings, and keys when you read double-bar graphs.

The graph in Example 1 is about immigration to the United States in two different years, 1985 and 1990. It has five pairs of bars. The labels under the pairs show that they stand for the places people emigrated from. The key shows that the left bar in each pair represents 1985 data. The right bar in each pair stands for 1990 data.

As Examples 1 and 2 show, studying a double-bar graph can help you make many observations based on the data it shows. Example 1 asks you to compare all the 1985 bars.

Example 1: According to the following graph, where did the largest group of immigrants come from in 1985?

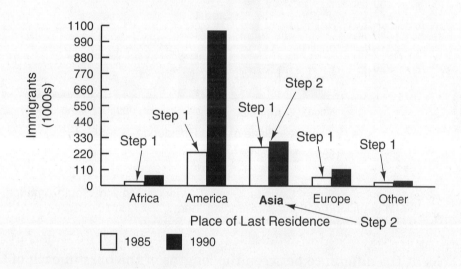

Note: "America" includes Canada, Central America, Mexico, South America, and the island nations surrounding them.
Source: U.S. Immigration and Naturalization Service.

STEP 1: Look at the key and find all the bars on the graph that stand for 1985.

STEP 2: Find the tallest of the 1985 bars. Look at the label under the pair it is part of. It tells you that the largest group of immigrants in 1985 came from Asia.

Example 2 asks you to compare the differences between the lengths of the bars in all five pairs on the graph.

Example 2: According to the following graph there were more immigrants from every part of the world in 1990 than in 1985. From which place was the increase in immigrants the greatest?

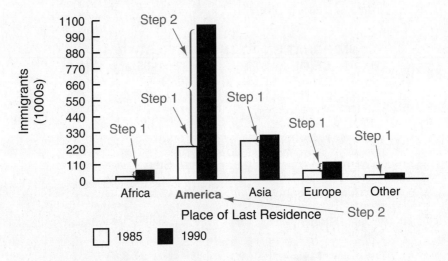

Note: "America" includes Canada, Central America, Mexico, South America, and the island nations surrounding them.
Source: U.S. Immigration and Naturalization Service.

STEP 1: Look at the difference between the lengths of the bars in each of the five pairs on the graph.

STEP 2: Find the pair with the greatest difference. Look at the label under that pair. It tells you that the greatest increase was in immigrants from America.

EXERCISE 43a

Answer the questions about the following double-bar graph.

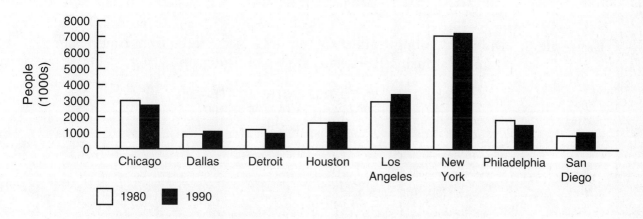

Source: April 1990 census, U.S. Bureau of the Census.

1. What kind of data about U.S. cities does the graph give?
2. For what two years are the data reported?
3. How many cities are included in the graph?
4. Which U.S. agency supplied the data for the graph?
5. What does the number *1000* on the scale stand for?
6. About how many people lived in the U.S. city that was the largest in 1990?
7. Which three cities' populations decreased between 1980 and 1990?
8. Which two cities' populations were less than one million in 1980?
9. Which city's population increased the most between 1980 and 1990?
10. Which city's population changed the least between 1980 and 1990?

Check your answers on page 244.

Double-Line Graphs

In Lesson 41 you learned that line graphs not only display data but also show trends. Double-line graphs have two lines, which makes it possible to compare the trends of two sets of data.

Usually the lines on a line graph are printed differently to distinguish between the sets of data they represent. For example, one line may be solid and the other dotted. The graph will have a key or use labels to explain the meaning of each line.

The graph in Example 3 is about recent immigration to the United States from two different places. It has two lines, each with three points on it. The key shows that the solid line represents immigration from countries other than the United States in America. The dotted line represents immigration from countries in Asia.

As Example 3 shows, studying a double-line graph can help you make observations based on the data and the trends it shows. Example 3 asks you to compare the points on a double-line graph or to compare the positions of the two lines.

Example 3: Did more immigrants come to the United States from America or from Asia during the period shown on the graph?

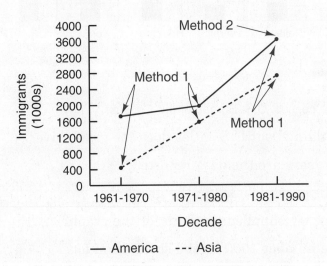

IMMIGRATION TO THE UNITED STATES FROM AMERICA AND ASIA, 1961–1990

Note: "America" includes Canada, Central America, Mexico, South America, and the island nations surrounding them.
Source: U.S. Immigration and Naturalization Service.

METHOD 1: For each decade, compare the points on both lines on the graph. With the help of the key, find which place is represented by the higher point in each pair.

METHOD 2: Find the line that is higher on the graph. With the help of the key, find which place is represented by that line.

Both methods show that more immigrants came to the United States from America than from Asia during each decade shown on the graph.

Comparing the angles of the lines on a double-line graph can give you useful information about trends.

Example 4: According to the following graph, from which place did the immigration trend rise sharply after 1980, America or Asia?

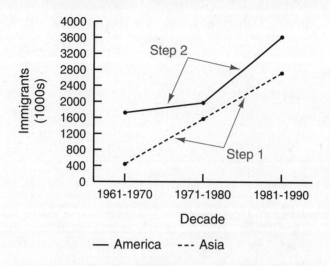

IMMIGRATION TO THE UNITED STATES FROM AMERICA AND ASIA, 1961–1990

Note: "America" includes Canada, Central America, Mexico, South America, and the island nations surrounding them.
Source: U.S. Immigration and Naturalization Service.

STEP 1: The rising line that stands for Asia on the graph is almost straight from 1961 to 1990. That means that immigration to the United States from Asia during that whole period rose at nearly a constant rate. In fact, each decade there were a little over 1,000,000 more immigrants than in the previous decade.

STEP 2: The line that stands for America on the graph rises more sharply after 1980 than after 1970. That means that the rate of immigration to the United States from America rose sharply after 1980. In fact, there were a little over 250,000 more immigrants from America in the 1970s than in the 1960s. In the 1980s there were well over 1,500,000 more American immigrants than in the 1970s. The increase in the 1980s was about six times the increase in the 1970s.

EXERCISE 43b

Answer the questions about the following double-line graph.

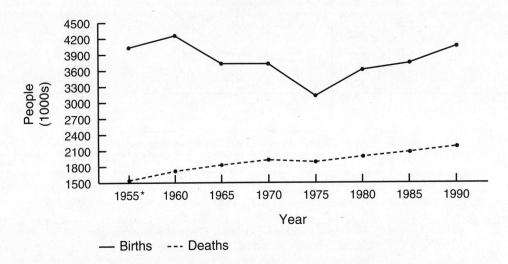

*Does not include Alaska and Hawaii.
Note: Beginning in 1970, nonresident births and deaths are excluded.
Source: National Center for Health Statistics, U.S. Department of Health and Human Services.

1. What kinds of data does the graph give?

2. The data reported cover what time period?

3. Which U.S. agency supplied the data for the graph?

4. What does the number *1500* on the scale stand for?

5. What is the increment between the numbers on the scale?

6. What kind of trend does the line that stands for deaths show: upward, level, or downward?

7. In which two years does the direction of the trend of the line that stands for births change?

8. After 1975 which increased more sharply, the number of births or the number of deaths?

9. About how many births were there in 1990?

10. About how many deaths were there in 1990?

11. In which year were the numbers of deaths and births closest to each other?

12. Between which two years reported did the number of births change least?

Check your answers on page 244.

Chapter 2

WORKING WITH GRAPHIC INFORMATION

The two lessons in this chapter introduce solving one-step and multistep problems based on tables, bar graphs, pictographs, line graphs, circle graphs, divided bar graphs, and double graphs. They will help you apply your whole-number, decimal, fraction, and percent skills in solving a variety of types of problems.

Solving One-Step Problems

As you learned in Lessons 40–43, graphic displays report data, show trends, illustrate the sizes of the parts of wholes, and make it possible to compare data. By applying math operations to the information in a graphic display, you can find facts that the display does not directly show. To find a fact, you may need to add, subtract, multiply, or divide data from the display.

Adding to Find Facts

You may want to find the sum of two or more of the values in a table or a graph. To do so, first read those values from the table or graph. Then, add.

POLITICAL DIVISIONS OF THE 90th to 102nd CONGRESSES

Congress	Years	U.S. Senate			U.S. House of Representatives			
		Democrats	Republicans	Other parties	Democrats	Republicans	Other parties	Vacant
90th	1967–69	64	36		248	187		
91st	1969–71	58	42		243	192		
92nd	1971–73	54	44	2	255	180		
93rd	1973–75	56	42	2	242	192	1	
94th	1975–77	61	37	2	291	144		
95th	1977–79	61	38	1	292	143		
96th	1979–81	58	41	1	277	158		
97th	1981–83	46	53	1	242	190		3
98th	1983–85	46	54		269	166		
99th	1985–87	47	53		253	182		
100th	1987–89	54	46		258	177		
101st	1989–91	57	43		262	173		
102nd*	1991–93	57	43		266	164	1	4

*As of May 10, 1991.
Source: Secretary of the Senate; Clerk of the House of Representatives.

Example 1: According to the table on page 197, what was the total membership of the House of Representatives in the 93rd Congress?

Step 1	Step 2
242 democrats	242 representatives
192 republicans	192 representatives
1 other parties	+ 1 representative
	435 representatives

STEP 1: In the *Congress* column find the row heading *93rd*. Look across that row to the four columns that show the members of the House of Representatives. Find the three values in those columns. (For the 93rd Congress, the *Vacant* column is empty.)

STEP 2: Using the three numbers, set up an addition problem and add. The total membership of the House of Representatives in the 93rd Congress was 435.

Subtracting to Find Facts

You may want to find the difference between two values in a graphic display. Read the values from the table or graph and subtract.

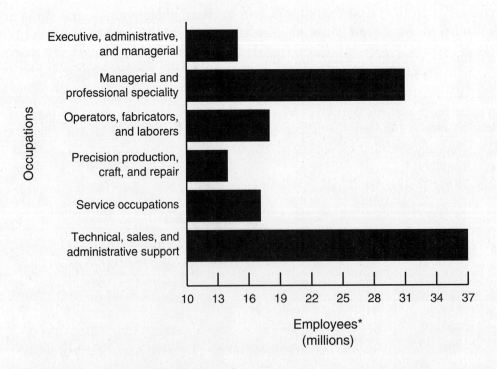

**U.S. EMPLOYMENT BY OCCUPATION
1990**

*Rounded to the nearest million.
Source: Bureau of Labor Statistics, U.S. Department of Labor.

198   UNIT 3: **Tables and Graphs**

Example 2: According to the graph on page 198, how many more people worked in managerial and professional specialty occupations than in service occupations in 1990?

Step 1	Step 2
Managerial and professional specialty = 31 million	31,000,000 employees
Service occupations = 17 million	− 17,000,000 employees
	14,000,000 employees

STEP 1: Read the values of the bars labeled *Managerial and professional specialty* and *Service occupations*.

STEP 2: Write the values in a column and subtract. 14,000,000 more people worked in managerial and professional specialty occupations than in service occupations in 1990.

EXERCISE 44a

Part A. Answer the questions about the following pictograph.

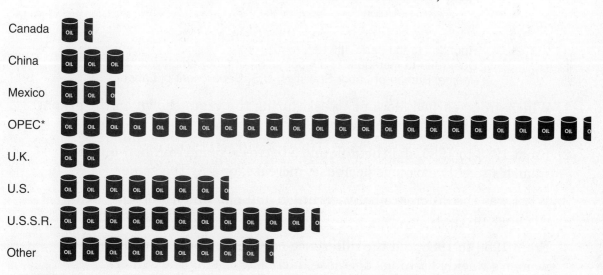

WORLDWIDE DAILY CRUDE OIL PRODUCTION, 1990

Key: [oil] = 1,000,000 barrels, rounded to the nearest half-million barrels.
*The Organization of the Petroleum Exporting Countries includes Algeria, Ecuador, Gabon, Indonesia, Iran, Iraq, Kuwait, Libya, Nigeria, Qatar, Saudi Arabia, United Arab Emirates, and Venezuela.
Source: Energy Information Administration, *Annual Energy Review, 1990.*

1. About how many barrels of crude oil did the OPEC nations produce per day in 1990?

2. About how many barrels of crude oil were produced in North America (Canada, Mexico and the United States) per day in 1990?

3. In 1990, about how many more barrels of crude oil did the U.S.S.R. produce per day than the United States?

4. Throughout the world, about how many barrels of crude oil were produced each day in 1990?

5. Which country produced more crude oil per day in 1990, China or Mexico? How much more?

Part B. Answer the questions about the following double-line graph.

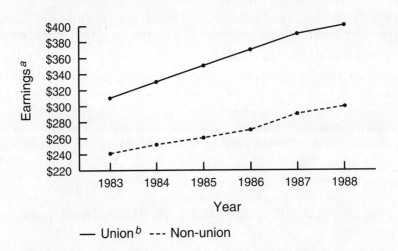

U.S. WOMEN'S MEDIAN WEEKLY EARNINGS

— Union[b] --- Non-union

[a] Rounded to the nearest $10.
[b] Data refer to members of a labor union or other employee association.
Source: Bureau of Labor Statistics, U.S. Department of Labor.

1. Which group earned more per week during the years shown on the graph, union women or non-union women?

2. From 1983 to 1986, did the difference between union and non-union women's weekly earnings decrease, increase, or stay the same?

3. What was the difference between union and non-union women's weekly earnings in 1984?

4. From 1986 to 1988, did the difference between union and non-union women's weekly earnings decrease, increase, or stay the same?

5. How great was that difference between union and non-union women's weekly earnings in 1988?

Check your answers on page 244.

Multiplying to Find Facts

To solve a personal-finance problem, a job-related problem, or some other kind of problem, you may want to use information you find in a graph. Many such problems can be solved by multiplication.

Example 3: Marion regulates her spending according to the following budget. How much does she spend on food each week if her take-home pay is $240?

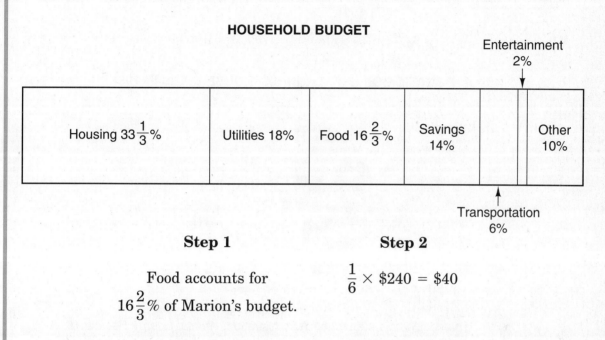

Step 1

Food accounts for $16\frac{2}{3}\%$ of Marion's budget.

Step 2

$\frac{1}{6} \times \$240 = \$40$

STEP 1: On the graph, find the percent of Marion's budget that she applies to food.

STEP 2: Multiply that percent times $240, Marion's take-home pay. Marion spends $40 a week on food.

Dividing to Find Facts

Using information you find in a graph, you can solve some problems by dividing.

Example 4: Marion wants to spend no more of her total income on rent than is shown in the divided-bar graph in Example 3. How much would her monthly income have to be for her to afford an apartment with a rent of $575?

Step 1

Housing accounts for $33\frac{1}{3}\%$ of the household budget.

Step 2

$\$575 \div \frac{1}{3} = \$1725$

Chapter 2: Working with Graphic Information

STEP 1: On the graph, find the percent of Marion's budget that she applies to housing.

STEP 2: Divide the rent by that percent. Marion's monthly income must be $1725 for her to afford a $575 apartment.

EXERCISE 44b

Answer the questions about the following circle graph.

APPLICATION OF ENERGY IN U.S. HOUSEHOLDS IN 1987

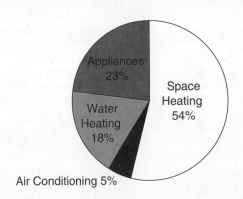

Note: Percents are rounded to the nearest 1%.
Source: Energy Information Administration, *Annual Energy Review 1990.*

1. Assume that an all-electric household uses energy in the proportions shown in the graph. If that household's annual electricity bill is $1800, how much does space heating cost each year?

2. Harry uses energy in the proportions shown in the graph and spends $599.94 a year on oil for space heating. What is his annual cost for energy?

3. Tanya was interested in knowing how much she spent each month to run her television and other appliances. If her energy use matches that shown in the graph and her monthly energy bill is $45, how much does Tanya pay for her applicances per month?

4. Latif lives in an all-electric apartment. Last year his bills for electricity totaled $480. If he had not run his air conditioner last summer, how much lower might his electricity charges have been? (Assume that Latif used energy in the proportions shown in the graph.)

5. According to the Energy Information Administration, Americans spent $17,595,000,000 on water heating in 1987. How much did Americans spend altogether on household energy in 1987?

Check your answers on page 245.

Solving Multistep Problems

Lesson 44 showed that you can apply math operations to the information in a graphic display to find facts that the display does not show directly. Sometimes you may need to perform two or more operations to find the fact you want. As in the following example, you may need to add, to subtract, and to multiply to solve a problem.

Example: A family's budget is shown in the following divided-bar graph. The family's income last month was $2000. How much did their budget allow them to spend in the category called *Other*?

FAMILY BUDGET

| Housing 25% | Utilities 22% | Food 20% | Savings 10% | Transportation 10% | Other ?% |

Step 1	Step 2	Step 3
Housing 25%	100%	$2000 × .13 = $260
Utilities 22%	− 87%	
Food 20%	Other 13%	
Savings 10%		
Transportation + 10%		
87%		

STEP 1: The percent of expenses called *Other* is not given in the graph. To find it, first add up all the percents that are given. They equal 87%.

STEP 2: Because the whole budget is 100%, subtract the total of the percents given from 100%. The difference is the percent in the *Other* category, 13%.

STEP 3: Multiply that percent times the family's monthly income: 13% of $2000 is $260.

Many different types of multistep problems can be solved using information from tables and graphs. The following exercise will give you practice at solving some of those types of problems.

EXERCISE 45

Answer the questions about the following graphic displays.

Questions 1 through 4 are based on the following table.

CALORIES USED PER MINUTE OF EXERCISE
by Exercise and Body Weight

Activity	Weight in Pounds					
	100	120	150	170	200	220
Bicycling (5.5 mph)	3.1	3.8	4.7	5.3	6.3	6.9
Bicycling (10 mph)	5.4	6.5	8.1	9.2	10.8	11.9
Calisthenics	3.3	3.9	4.9	5.6	6.6	7.2
Jogging (5.5 mph)	6.1	7.3	9.1	10.4	12.2	13.4
Running (7.5 mph)	9.4	11.3	14.1	16.0	18.8	20.7
Swimming	4.8	5.7	7.2	8.1	9.6	10.5
Volleyball	2.3	2.7	3.4	3.9	4.6	5.0
Walking (3 mph)	2.7	3.2	4.0	4.6	5.4	5.9
Walking (4 mph)	3.9	4.6	5.8	6.6	7.8	8.5

Note: In addition to body weight, many other factors (including air temperature, clothing, and the vigor with which a person exercises) can affect the number of calories used.

1. How many calories would a 150-pound person use bicycling at 5.5 mph for 20 minutes and then playing volleyball for 15 minutes?

2. Myrna weighs 120 pounds. She ran at 7.5 mph for 10 minutes, swam for 10 minutes, and then did calisthenics for 10 minutes. How many calories did she use doing those three exercises?

3. Two men played volleyball with their teams for $\frac{1}{2}$ hour. One of the men weighs 220 pounds; the other weighs 170 pounds. How many more calories did the heavier man use than the lighter man?

4. Frank weighs 200 pounds. He took in 195 calories more than his diet called for one day and decided to try to walk off the excess. How many minutes would he have to walk at 4 mph to use up the excess calories?

Questions 5 and 6 are based on the following pictograph.

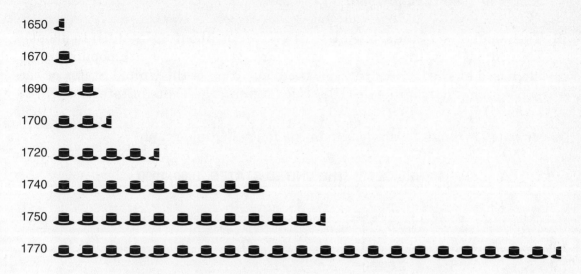

5. By what percent did the population of the American colonies increase between 1670 and 1770?

6. During which twenty-year period did the population of the American colonies increase by less than 100%, 1700–1720 or 1720–1740?

Questions 7 through 9 are based on the following double-bar graph.

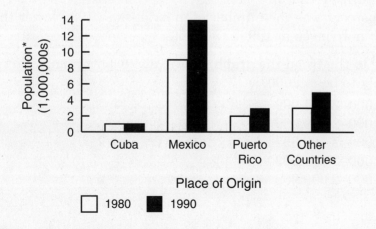

*Rounded to the nearest million.
Source: Bureau of the Census, U.S. Department of Commerce.

7. By what percent did the population of immigrants from Mexico increase between 1980 and 1990? (Round your answer to the nearest .1%.)

8. How many more immigrants of Hispanic origin were there in the United States in 1990 than in 1980?

9. The population of the United States grew from about 227 million to about 249 million between 1980 and 1990. During that time, which population increased by the greater percent, the population of the United States or the population of immigrants of Hispanic origin in the United States?

Questions 10 through 12 are based on the following line graph.

MARRIAGES IN THE UNITED STATES, 1960–1990

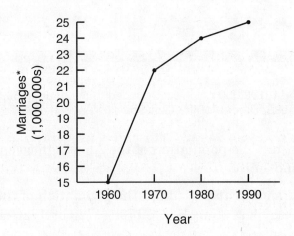

*Rounded to the nearest 100,000.
Source: National Center for Health Statistics.

10. If the number of marriages increased steadily between 1970 and 1980, about how many marriages were there in 1975?

11. By what percent was the number of marriages in 1980 lower than the number of marriages in 1990?

12. According to the trend the graph shows, about how many marriages are there likely to be in 2000?
    (1) 2,600,000–2,699,999
    (2) 2,500,000–2,599,999
    (3) 2,400,000–2,499,999
    (4) 2,300,000–2,399,999
    (5) 2,000,000–2,099,999

Questions 13 through 15 are based on the following circle graph.

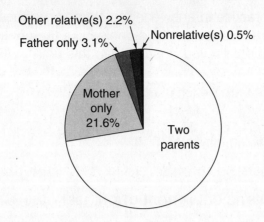

LIVING ARRANGEMENTS IN THE UNITED STATES IN 1990
64 Million Children Under 18

Note: Excludes people under 18 who head households.
Source: U.S. Bureau of the Census.

13. What percent of children under 18 lived with both their parents in 1990?

14. About how many children lived with only one parent in 1990?

15. About how many children lived with neither of their parents in 1990?

Check your answers on page 245.

GED PRACTICE 3

This section will give you practice in answering questions like those on the GED. The 10 questions in this Practice are multiple-choice like the ones on the GED. As you do this Practice, use the skills you've studied in this unit.

Directions: Choose the one best answer to each item.

Items 1 through 4 refer to the following table.

U.S. DOMESTIC COIN PRODUCTION, 1981–1990

Year	Pennies	Nickels	Dimes	Quarters	Halves	Total
1981	12,864,985,677	1,022,305,843	1,388,934,143	1,177,438,833	57,383,533	16,511,048,029
1982	16,725,504,368	666,081,544	1,062,188,584	980,973,788	23,959,102	19,458,707,386
1983	14,219,554,428	1,098,341,276	1,377,154,224	1,291,341,446	66,611,244	18,053,002,618
1984	13,720,317,906	1,264,444,146	1,561,472,976	1,223,028,064	52,291,158	17,821,554,250
1985	10,935,889,813	1,106,862,408	1,293,180,932	1,295,781,850	38,520,996	14,670,235,999
1986	8,934,262,191	898,702,633	1,155,976,667	1,055,497,993	28,473,778	12,072,913,262
1987	9,561,856,445	782,090,085	1,415,912,883	1,238,094,177	99,481,000	13,097,434,590
1988	11,346,550,443	1,435,131,652	1,992,935,488	1,158,862,687	25,626,096	15,959,106,366
1989	12,837,140,268	1,497,523,652	2,240,355,488	1,417,290,422	41,196,188	18,033,506,018
1990	12,031,422,711	1,415,222,474	1,956,105,597	1,560,357,858	43,614,192	17,006,722,832
Total	123,177,484,250	11,186,705,713	15,444,216,982	12,398,667,118	477,157,287	162,684,231,350

Source: U.S. Mint, U.S. Department of the Treasury.

1. How many quarters were minted in the United States in 1983?

    (1) 18,053,022,618
    (2) 12,398,667,118
    (3) 1,377,154,224
    (4) 1,291,341,446
    (5) 980,973,788

2. What was the value of the dimes minted in the United States from 1981 through 1990?

    (1) $ 15,444,216.98
    (2) $ 154,442,169.82
    (3) $ 1,544,421,698.20
    (4) $ 15,444,216,982.00
    (5) $ 154,442,169,820.00

3. During each year which type of coin was the most minted?

    (1) pennies
    (2) nickels
    (3) dimes
    (4) quarters
    (5) halves

4. About 10% of all the coins produced from 1981 through 1990 were minted in

    (1) 1981
    (2) 1983
    (3) 1985
    (4) 1987
    (5) 1989

208   UNIT 3: Tables and Graphs

Items 5 and 6 refer to the following graph.

U.S. AUTOMOBILE SALES IN 1985 AND 1990

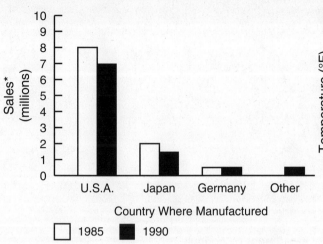

*Rounded to the nearest half million.
Note: Amount for "other" in 1985 was less than a quarter million.
Source: Motor Vehicle Manufacturers Association.

5. About how many automobiles were sold in the United States in 1990?

   (1) 10,500,000
   (2) 9,500,000
   (3) 8,000,000
   (4) 7,000,000
   (5) Not enough information is given.

6. Compared to 1985, 1990 U.S. sales of Japanese-made cars

   (1) decreased by 25%
   (2) increased by 25%
   (3) decreased by 75%
   (4) increased by 75%
   (5) decreased by $133\frac{1}{3}$%

Items 7 and 8 refer to the following graph.

MONTHLY NORMAL TEMPERATURE IN CHICAGO, ILLINOIS

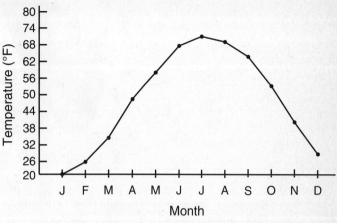

Source: National Climatic Data Center, NESDIS, NOAA, U.S. Department of Commerce.

7. The difference between Chicago's highest and lowest monthly normal temperatures is roughly

   (1) 10°
   (2) 20°
   (3) 30°
   (4) 40°
   (5) 50°

8. The trend of Chicago's monthly normal temperatures changes after both

   (1) January and July
   (2) January and June
   (3) December and January
   (4) December and July
   (5) December and June

Items 9 and 10 refer to the following graph.

LIVESTOCK ON U.S. FARMS IN 1991

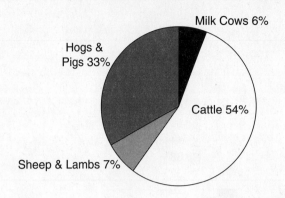

Note: Based on an animal population of 180 million, rounded to the nearest 10 million. Data are accurate to the nearest 1%.
Source: National Agricultural Statistics Service: U.S. Department of Agriculture.

9. Roughly how many hogs and pigs were in the 1991 U.S. livestock population?
    (1)   33,000,000
    (2)   60,000,000
    (3)   180,000,000
    (4)   550,000,000
    (5)   Not enough information is given.

10. Sheep and lambs were valued at an average of $65.60 per head in 1991. The total value of the sheep and lamb population that year was about
    (1) $11,808,000,000
    (2) $ 826,560,000
    (3) $ 180,000,000
    (4) $ 12,600,000
    (5) $ 11,808,000

Check your answers on page 247.

GED PRACTICE 3 SKILLS CHART

To review the mathematics skills covered by the items in GED Practice 3, study the following lessons in Unit 3.

Unit 3	Tables and Graphs	Item Number
Lesson 40	Graphic Displays of Data	1, (2), 3, (4–6)
Lesson 41	Graphic Displays of Trends	(7), 8
Lesson 42	Graphic Displays of the Parts of Wholes	(9, 10)
Lesson 43	Double Graphs	(5–6)
Lesson 44	Solving One-Step Problems	2, 4, 5, 7, 9
Lesson 45	Solving Multistep Problems	6, 10

Posttest

This posttest will help you review the work you've done in this book. It will also give you practice in answering questions like those on the GED. It is made up of 28 multiple-choice questions, half as many questions as there are on the Mathematics Test of the GED. The skills chart on page 218 will help you plan a review of the skills covered by this book that you may need to practice further.

MATHEMATICS POSTTEST

Directions: Choose the <u>one best answer</u> to each item.

1. In the precision tool shop where he works Tom laid out these five drill bits in order of size from smallest diameter to largest diameter:

    A. $\frac{25}{32}$ in. B. $\frac{13}{16}$ in. C. $\frac{5}{8}$ in. D. $\frac{3}{4}$ in. E. $\frac{1}{2}$ in.

    In which order did he lay out the bits?
    (1) E, D, C, B, A
    (2) E, C, D, A, B
    (3) C, B, E, D, A
    (4) B, A, D, C, E
    (5) A, B, C, D, E

2. The first 3 ingredients in a recipe for pound cake are

    $1\frac{1}{3}$ C butter
    $2\frac{1}{3}$ C flour
    $1\frac{2}{3}$ C sugar

    After those ingredients are mixed together, how many cups of batter will there be?
    (1) $4\frac{1}{3}$
    (2) 5
    (3) $5\frac{1}{3}$
    (4) $8\frac{1}{3}$
    (5) $14\frac{2}{3}$

3. In the first week of his exercise program, Willie jogged each day before he went to work. These are the distances he ran:

    Monday $\frac{1}{10}$ mi
    Tuesday $\frac{1}{4}$ mi
    Wednesday $\frac{2}{5}$ mi
    Thursday $\frac{1}{2}$ mi
    Friday $\frac{3}{4}$ mi

    How many miles in all did he run that week?
    (1) $\frac{8}{25}$
    (2) $\frac{2}{5}$
    (3) $\frac{1}{2}$
    (4) 2
    (5) 40

4. At the start of a trip Lamar had $12\frac{3}{4}$ gallons of gas in her car. At the end of the trip she had $7\frac{1}{3}$ gallons left. How many gallons of gas did her car use during the trip?
    (1) $1\frac{65}{88}$
    (2) $5\frac{5}{12}$
    (3) $5\frac{1}{2}$
    (4) $20\frac{1}{2}$
    (5) $93\frac{1}{2}$

5. Alan cut a $17\frac{1}{4}$-inch board into two pieces. One piece was $9\frac{7}{8}$ inches long. How many inches long was the other piece?
    (1) $1\frac{59}{79}$
    (2) $7\frac{3}{8}$
    (3) $8\frac{5}{8}$
    (4) $27\frac{1}{8}$
    (5) $170\frac{11}{32}$

6. Cassie divides her 40-hour work week among 3 projects. She earns $325 per week. Since the Learning Center project takes $\frac{3}{5}$ of her time, how much of her weekly wage does she earn from that project?
    (1) $ 8.13
    (2) $ 65.00
    (3) $108.33
    (4) $195.00
    (5) $541.67

7. In a bolt of printed cloth, a large pattern is repeated $2\frac{2}{5}$ times each yard. How many times will the pattern appear on a piece of cloth $6\frac{1}{4}$ yards long?
    (1) $2\frac{29}{48}$
    (2) $3\frac{17}{20}$
    (3) $6\frac{1}{4}$
    (4) $8\frac{13}{20}$
    (5) 15

8. Jack and Maria invited 4 other couples to dinner. Jack used $2\frac{1}{2}$ pounds of ground beef to make 2-ounce meatballs. When he served the dinner, he put an equal number of meatballs on each plate. How many meatballs did each person get?

   (1) 2
   (2) $2\frac{1}{2}$
   (3) $3\frac{3}{4}$
   (4) 5
   (5) 20

Items 9 and 10 refer to the following drawing.

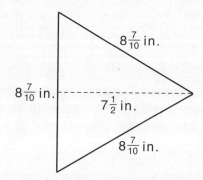

9. Tamika is making flags like the one shown for the members of her club to carry in a parade. To strengthen the flags, Tamika wants to sew a thin strip of canvas around all three edges of each flag. How many inches of canvas will she need for each flag?

   (1) $8\frac{7}{10}$
   (2) $13\frac{1}{20}$
   (3) $17\frac{2}{5}$
   (4) $26\frac{1}{10}$
   (5) $33\frac{3}{5}$

10. To figure out how much material she needs to make a flag like the one shown, Tamika calculated (in square inches) the area of one flag to be

    (1) $26\frac{1}{10}$
    (2) $32\frac{5}{8}$
    (3) $37\frac{169}{200}$
    (4) $65\frac{1}{4}$
    (5) $75\frac{69}{100}$

Item 11 refers to the following figure.

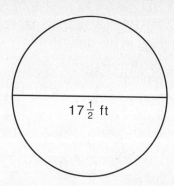

11. Ramon wants to plant rose bushes around the edge of this circular garden. He plans to place the bushes $2\frac{1}{2}$ feet apart. How many rose bushes will he need?

    (1) 22
    (2) 55
    (3) $96\frac{1}{4}$
    (4) $137\frac{1}{2}$
    (5) $240\frac{5}{8}$

Item 12 refers to the following figure.

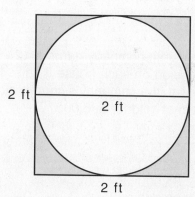

12. Tan Hock is planning to cover a floor with small gray and white ceramic tiles. The figure shows the basic pattern he will repeat over and over on the floor. The unshaded portion of the pattern stands for the white tiles; the shaded portion, for the gray tiles. Tan Hock needs to figure out how many gray tiles to buy. To begin his figuring, he correctly calculates the area (in square feet) of the shaded portion of one pattern to be

    (1) $\frac{6}{7}$
    (2) $2\frac{2}{7}$
    (3) $3\frac{1}{7}$
    (4) 4
    (5) $7\frac{1}{7}$

Posttest   213

13. Manuel installed ceiling tiles in his bedroom. He used $10\frac{2}{3}$ boxes of tiles. Each box cost him $12.75. What was the total cost for the tiles he used in his bedroom?
    (1) $ 1.20
    (2) $ 2.08
    (3) $ 23.42
    (4) $127.50
    (5) $136.00

14. Which of the following does not have the same value as the others?
    (1) .2
    (2) $\frac{2}{10}$
    (3) $\frac{1}{5}$
    (4) 20%
    (5) 2%

15. For his store to make a profit, Todd marks each item up by $\frac{3}{8}$ over his cost to arrive at its selling price. His percent of mark-up is
    (1) .00375
    (2) .375
    (3) 3.75
    (4) 37.5
    (5) 375

16. A real estate company earns a $12\frac{1}{2}$% commission on each house it sells. How much was the company's commission on a house that sold for $96,000?
    (1) $ 7,680
    (2) $ 12,000
    (3) $ 12,500
    (4) $ 84,000
    (5) $108,000

17. On sale a $350-suit cost only $245. What percent of the original price was the sale price?
    (1) 7%
    (2) 30%
    (3) 42.9%
    (4) 70%
    (5) 142.9%

18. Henrique sells storm windows. He earns a 4.5% commission on his total sales each week. Last week his commission was $157.50. What were Henrique's sales last week?
    (1) $ 7.09
    (2) $ 35.00
    (3) $ 164.59
    (4) $ 350.00
    (5) $3500.00

19. Angie wanted to buy a small TV for her bedroom. She saw a $79.98 set on sale for 20% off. About how much did the TV cost on sale?
    (1) $ 16
    (2) $ 64
    (3) $ 72
    (4) $ 80
    (5) $100

20. Janice drives 400 miles to visit her mother. She always makes it a 2-day trip so she doesn't arrive tired. She drives 350 miles the first day. What percent of her trip does she drive the second day?
    (1) $12\frac{1}{2}$%
    (2) $16\frac{2}{3}$%
    (3) $62\frac{1}{2}$%
    (4) $83\frac{1}{3}$%
    (5) $87\frac{1}{2}$%

21. A bank advertised a $5\frac{1}{2}$% annual interest rate for savings accounts. If you made one deposit of $1200 in such an account, what would your account balance be at the end of a year?
    (1) $ 66
    (2) $1134
    (3) $1200
    (4) $1266
    (5) $6600

22. When Sarah began her Pre-GED course, 30 people were enrolled. After six weeks, 20% of the original class had dropped out. How many students were still taking the course after six weeks?

    (1) 2
    (2) 6
    (3) 24
    (4) 28
    (5) 36

23. A deli pays $1.50 for a half-gallon of milk and sells it for $1.56. By what percent does the deli mark up the price of the milk?

    (1) 3.8%
    (2) 4.0%
    (3) 6.0%
    (4) 96.2%
    (5) 104.0%

Item 24 refers to the following graph.

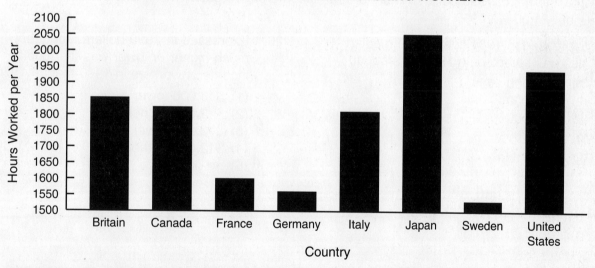

Reprinted with permission of the *Democrat and Chronicle,* Rochester, N.Y., 1992.

24. About how many fewer hours did German manufacturing workers work in 1990 than manufacturing workers in the United States?

    (1) 3500
    (2) 1950
    (3) 1500
    (4) 500
    (5) 400

Items 25 and 26 refer to the following graph and information.

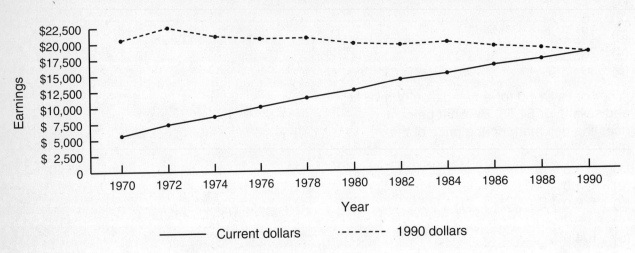

Source: Bureau of Labor Statistics, U.S. Department of Labor.

This graph reports median annual earnings in two ways.

The line labeled *Current dollars* reports earnings in terms of that year's dollars. For example, the 1977 median income was $10,000 and was worth $10,000 in terms of 1977 dollars.

The line labeled *1990 dollars* adjusts for inflation and other factors. It reports each year's median earnings in terms of 1990 dollars. Therefore, an income of $10,000 in 1977 was comparable to an income of $21,000 in 1990.

25. Expressed in 1984 dollars, the median income of workers in 1984 was about

    (1) $14,000
    (2) $15,000
    (3) $16,000
    (4) $19,000
    (5) $19,500

26. Expressed in 1990 dollars, about how much higher or lower was the median income of workers in 1990 than in 1970?

    (1) $ 3,000 lower
    (2) $ 3,000 higher
    (3) $12,000 lower
    (4) $12,000 higher
    (5) no difference

Items 27 and 28 refer to the following graph and information.

U.S. FAMILY INCOME DISTRIBUTION, 1990
Total Personal Income: $4645.5 Billion

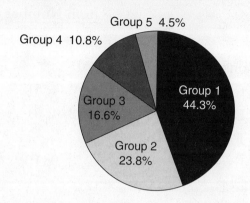

Sources: Bureau of the Census and Bureau of Economic Analysis, U.S. Department of Commerce.

This graph shows how all personal income was spread among all the families in the United States in 1990.

The graph divides all U.S. families into five groups of equal size. Therefore, $\frac{1}{5}$ of all families are in Group 1, $\frac{1}{5}$ are in Group 2, etc.

The personal incomes of the families in each group differ: families in Group 1 have the highest incomes; families in Group 5 have the lowest incomes.

27. In 1990 nearly half of all personal income was received by families in

    (1) Group 1
    (2) Group 2
    (3) Group 3
    (4) Group 4
    (5) Group 5

28. Compared to the families in Group 5, how much more personal income did the families in Group 1 receive in 1990?

    (1) $4645.5 billion
    (2) $2057.9565 billion
    (3) $1848.918 billion
    (4) $ 464.55 billion
    (5) $ 209.0385 billion

Check your answers on page 247.

MATHEMATICS POSTTEST SKILLS CHART

To review the mathematics skills covered by the items in the Posttest, study at least the following lessons.

Unit 1	Fractions	Item Number
Lesson 1	Proper Fractions	All fraction problems
Lesson 2	Simplifying Proper Fractions	(20)
Lesson 3	Raising Proper Fractions to Higher Terms	(1, 3, 4, 5)
Lesson 4	Finding Lowest Common Denominators	(1, 3, 4, 5)
Lesson 5	Comparing and Ordering Fractions	1
Lesson 6	Improper Fractions and Mixed Numbers	(2, 3, 9, 10, 11)
Lesson 7	Adding Fractions with Like Denominators	(2)
Lesson 8	Adding Mixed Numbers with Like Denominators	2
Lesson 9	Adding Fractions with Unlike Denominators	3
Lesson 12	Subtracting with Unlike Denominators	4, (5)
Lesson 13	Borrowing	5, (12)
Lesson 14	Multiplying Fractions by Fractions	(6, 7, 8, 9, 10, 11, 13)
Lesson 15	Canceling before Multiplying Fractions	(6, 7, 8, 11, 13)
Lesson 16	Multiplying Whole Numbers by Fractions	6
Lesson 17	Multiplying Mixed Numbers by Fractions or Mixed Numbers	7, (9, 10, 11, 13)
Lesson 18	Dividing by Fractions	(8)
Lesson 19	Dividing Fractions or Mixed Numbers by Whole Numbers	(10)
Lesson 20	Dividing by Mixed Numbers	(11)
Lesson 21	Working with Measurements Expressed as Fractions or Mixed Numbers	8
Lesson 22	Finding Perimeters, Circumferences, and Areas	9, 10, 11, 12
Lesson 23	Working with Both Fractions and Decimals	13
Lesson 25	Solving Word Problems with Fractions and Mixed Numbers	(1, 2, 3, 4, 5, 6, 7, 9, 10, 13)
Lesson 26	Solving Multistep Word Problems with Fractions and Mixed-Numbers	(8, 11, 12)

Unit 2 Percents	Item Number
Lesson 27 Understanding Percents	All percent problems
Lesson 28 Rewriting Percents as Decimals	(14, 16, 18, 19, 21, 22)
Lesson 29 Rewriting Decimals as Percents	(14, 17, 23)
Lesson 30 Rewriting Percents as Fractions	(14, 22)
Lesson 31 Rewriting Fractions as Percents	(14, 15, 20)
Lesson 32 Recognizing Common Equivalents	14, 15, (20, 22)
Lesson 33 Understanding the Percent Triangle	(16, 17, 18, 19, 20, 21, 22, 23)
Lesson 34 Finding the Part	16, (19), 21, 22
Lesson 35 Finding the Percent	17, 20, 23
Lesson 36 Finding the Whole	18
Lesson 37 Estimating Solutions to Percent Problems	19
Lesson 38 Solving One-Step Percent Word Problems	(16, 17, 18, 19)
Lesson 39 Solving Multistep Percent Problems	(20, 21, 22, 23)

Unit 3 Tables and Graphs	
Lesson 40 Graphic Displays of Data	(24)
Lesson 41 Graphic Displays of Trends	25, (26)
Lesson 42 Graphic Displays of the Parts of Wholes	27, (28)
Lesson 43 Double Graphs	(25, 26)
Lesson 44 Solving One-Step Problems	24, 26
Lesson 45 Solving Multistep Problems	28

Answers and Solutions

In this section are the answers for all the problems in this book. To help you check your work, the solutions for many problems are shown.

PRETEST (page 1)

1. $\frac{3}{4}$
2. $\frac{3}{4}$
3. $\frac{1}{4}$
4. $\frac{12}{16}$
5. $\frac{18}{24}$ $\frac{20}{24}$ $\frac{14}{24}$
6. $\frac{7}{10}$
7. $\frac{4}{9}$ $\frac{7}{15}$ $\frac{3}{5}$ $\frac{2}{3}$
8. $3\frac{5}{8}$
9. $\frac{12}{5}$
10. $1\frac{3}{7}$
11. $8\frac{1}{4}$
12. $1\frac{1}{3}$
13. $10\frac{7}{24}$
14. $\frac{1}{3}$
15. $1\frac{1}{5}$
16. $\frac{5}{12}$
17. $7\frac{5}{12}$
18. $\frac{2}{15}$
19. $\frac{2}{3}$
20. $3\frac{6}{7}$
21. **12**
22. $5\frac{1}{2}$
23. $1\frac{1}{4}$
24. $1\frac{1}{2}$
25. $1\frac{2}{3}$ **yd**
26. **12 pt**
27. $8\frac{1}{2}$ **ft**
28. $4\frac{1}{8}$ **sq ft**
29. $1\frac{1}{2}$ **in.**
30. $\frac{3}{32}$ **sq in.**
31. **4 mi**
32. $1\frac{3}{11}$ **sq mi**
33. **.875**
34. $\frac{3}{20}$
35. $1\frac{3}{8}$ **lb** or **1.375 lb**
36. **(2)**
37. Hermiña used $2\frac{1}{8}$ **cups** of sugar in each batch.
    $6\frac{3}{8}$ C ÷ 3 = $2\frac{1}{8}$ C
38. Gerta will need **45 lb** of yarn.
    $6\frac{3}{4}$ ft × 8 ft = 54 sq ft
    54 sq ft × $1\frac{1}{6}$ lb per sq ft = 63 lb

39. **26%**
40. **135%**
41. **.08**
42. **62.5%** or **$62\frac{1}{2}$%**
43. $\frac{7}{8}$
44. **37.5%** or **$37\frac{1}{2}$%**
45. (a) **$400** is the whole.
    (b) **20%** is the percent.
    (c) **$80** is the part.
46. **70**
47. **45%**
48. **$550**
49. **(3)**
    $33\frac{1}{3}$% = $\frac{1}{3}$
    $2970 can be rounded to $3000.
    $\frac{1}{3}$ × $3000 = $1000
50. Jerry paid **$13.50 per sq yd.**
    $15 per sq yd × .9 = $13.50
51. Marty did **$17,100** in business this year.
    100% + 14% = 114%
    1.14 × $15,000 = $17,100
52. **Lau** sold the most subscriptions to *Worldwide Magazine* in 1992.
53. Martinez sold **7 million** subscriptions to *Worldwide Magazine* in 1992.
54. In 1992, Williams sold **2 million** more subscriptions to *Worldwide Magazine* than Thomas.
    6 million − 4 million = 2 million
55. Lau's *Worldwide Magazine* subscription sales were **29%** of all the subscription sales in 1992.
    5 million + 9 million + 7 million
    + 4 million + 6 million = 31 million
    9 million ÷ 31 million = .29 (rounded) = 29%

56. On average, Canandaigua Lake's water level is **688.5 feet above sea level** in June.
57. On average, Canandaigua Lake's water level rises most sharply between **March** and **April**.
58. In 1990, Canandaigua Lake's water level was the same as or lower than the average during the four months from **July through October.**
59. Canandaigua Lake's water level was **1 foot** higher in April, 1990 than in an average April.
    689.5 ft − 688.5 ft = 1 ft
60. Sales of **electrical equipment** accounted for the least business in Miller's Variety Store in 1992.
61. Hardware sales accounted for **25%** of all sales in Miller's Variety Store in 1992.
62. The dollar value of hardware sales in Miller's Variety Store in 1992 was **$80,000.**
    $320,000 × $\frac{1}{4}$ (or .25) = $80,000
63. The dollar value of sales of photo equipment and sporting goods together in Miller's Variety Store in 1992 was **$192,000.**
    15% + 45% = 60%
    $320,000 × .6 = $192,000

UNIT 1 FRACTIONS

EXERCISE 1 (PAGE 11)

Part A

1. (a) $\frac{1}{4}$ 2. (a) $\frac{3}{8}$
   (b) $\frac{3}{4}$ (b) $\frac{5}{8}$
3. (a) $\frac{5}{6}$ 4. (a) $\frac{2}{5}$
   (b) $\frac{1}{6}$ (b) $\frac{3}{5}$
5. (a) $\frac{2}{3}$ 6. (a) $\frac{1}{12}$
   (b) $\frac{1}{3}$ (b) $\frac{11}{12}$
7. (a) $\frac{1}{2}$ 8. (a) $\frac{5}{8}$
   (b) $\frac{1}{2}$ (b) $\frac{3}{8}$
9. (a) $\frac{5}{6}$ 10. (a) $\frac{1}{8}$
   (b) $\frac{1}{6}$ (b) $\frac{7}{8}$
11. (a) $\frac{4}{9}$ 12. (a) $\frac{7}{12}$
    (b) $\frac{5}{9}$ (b) $\frac{5}{12}$

Part B

Note: Your answer is correct if you shaded the correct number of parts. You need not have shaded exactly the same parts as shown below.

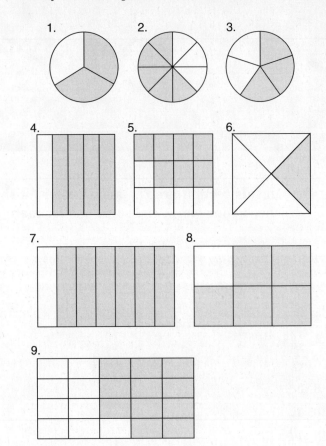

Word Problem

Sandy spent $\frac{3}{8}$ of Monday on the report. Monday had 8 parts, or hours, so 8 is the denominator. Sandy spent 3 of those parts, or hours, on the report, so 3 is the numerator.

EXERCISE 2A (PAGE 14)

1. $\frac{5}{7}$ 2. $\frac{1}{12}$ 3. $\frac{3}{5}$ 4. $\frac{3}{4}$
5. $\frac{1}{2}$ 6. $\frac{2}{5}$ 7. $\frac{4}{5}$ 8. $\frac{2}{3}$
9. $\frac{2}{5}$ 10. $\frac{9}{20}$

Word Problem

The baker used $\frac{3}{5}$ of the flour.

$$\frac{6 \div 2}{10 \div 2} = \frac{3}{5}$$

EXERCISE 2B (PAGE 15)

1. $\frac{2}{3}$ 2. $\frac{1}{2}$ 3. $\frac{2}{3}$ 4. $\frac{5}{12}$
5. $\frac{4}{7}$ 6. $\frac{7}{8}$ 7. $\frac{1}{3}$ 8. $\frac{4}{5}$
9. $\frac{3}{5}$ 10. $\frac{3}{5}$ 11. $\frac{3}{10}$ 12. $\frac{1}{6}$
13. $\frac{5}{6}$ 14. $\frac{2}{5}$ 15. $\frac{1}{9}$ 16. $\frac{2}{5}$
17. $\frac{1}{2}$ 18. $\frac{4}{9}$ 19. $\frac{4}{11}$ 20. $\frac{2}{5}$
21. $\frac{5}{8}$ 22. $\frac{8}{9}$ 23. $\frac{1}{3}$ 24. $\frac{6}{11}$
25. $\frac{1}{4}$ 26. $\frac{2}{9}$ 27. $\frac{1}{3}$ 28. $\frac{2}{7}$
29. $\frac{2}{3}$ 30. $\frac{3}{20}$

Word Problem

Kwok Eng has already sold $\frac{1}{3}$ of the book.

$$\frac{8 \div 8}{24 \div 8} = \frac{1}{3}$$

Note: To reduce this fraction, you could divide each number by 2 and then by 4.

$$\frac{8 \div 2}{24 \div 2} = \frac{4 \div 4}{12 \div 4} = \frac{1}{3}$$

You could also divide each number by 4 and then 2.

$$\frac{8 \div 4}{24 \div 4} = \frac{2 \div 2}{6 \div 2} = \frac{1}{3}$$

EXERCISE 3A (PAGE 17)

Part A

1. $\frac{14}{16}$ 2. $\frac{3}{12}$
3. $\frac{8}{12}$ 4. $\frac{10}{25}$
5. $\frac{6}{60}$

Part B

Sample answers. For these sample answers, fractions were raised to higher terms by multiplying each by 2, by 3, and by 4. If you multiplied by higher numbers, your answers will be different.

1. $\frac{2}{4}$ $\frac{3}{6}$ $\frac{4}{8}$ 7. $\frac{2}{12}$ $\frac{3}{18}$ $\frac{4}{24}$
2. $\frac{6}{8}$ $\frac{9}{12}$ $\frac{12}{16}$ 8. $\frac{8}{10}$ $\frac{12}{15}$ $\frac{32}{40}$
3. $\frac{10}{12}$ $\frac{15}{18}$ $\frac{20}{24}$ 9. $\frac{10}{16}$ $\frac{15}{24}$ $\frac{20}{32}$
4. $\frac{6}{16}$ $\frac{9}{24}$ $\frac{12}{32}$ 10. $\frac{6}{20}$ $\frac{9}{30}$ $\frac{12}{40}$
5. $\frac{10}{18}$ $\frac{15}{27}$ $\frac{20}{36}$ 11. $\frac{4}{14}$ $\frac{6}{21}$ $\frac{8}{28}$
6. $\frac{2}{6}$ $\frac{3}{9}$ $\frac{4}{12}$ 12. $\frac{6}{24}$ $\frac{9}{36}$ $\frac{12}{48}$

EXERCISE 3B (PAGE 19)

Part A

1. $\frac{4}{20}$ 2. $\frac{6}{36}$ 3. $\frac{3}{30}$
4. $\frac{6}{18}$ 5. $\frac{9}{45}$ 6. $\frac{15}{24}$
7. $\frac{3}{9}$ 8. $\frac{6}{9}$ 9. $\frac{6}{8}$
10. $\frac{70}{100}$

Part B

1. $\frac{4}{10}$ 2. $\frac{3}{36}$ 3. $\frac{28}{40}$
4. $\frac{6}{14}$ 5. $\frac{24}{30}$ 6. $\frac{22}{32}$
7. $\frac{15}{35}$ 8. $\frac{32}{36}$ 9. $\frac{6}{10}$
10. $\frac{9}{24}$

Word Problem

Hannah needs $\frac{5}{10}$ gallon of gasoline.

$$\frac{1}{2} = \frac{?}{10}$$

$$\frac{1 \times 5}{2 \times 5} = \frac{5}{10}$$

EXERCISE 4A (PAGE 21)

1. $\frac{4}{24}$ and $\frac{9}{24}$ 2. $\frac{5}{20}$ and $\frac{14}{20}$ 3. $\frac{5}{40}$ and $\frac{4}{40}$
4. $\frac{6}{60}$ and $\frac{5}{60}$ 5. $\frac{3}{12}$ and $\frac{10}{12}$ 6. $\frac{25}{30}$ and $\frac{8}{30}$
7. $\frac{9}{18}, \frac{6}{18}$, and $\frac{2}{18}$ 8. $\frac{6}{12}, \frac{8}{12}$, and $\frac{3}{12}$
9. $\frac{3}{12}, \frac{8}{12}$, and $\frac{2}{12}$ 10. $\frac{25}{40}, \frac{12}{40}$, and $\frac{30}{40}$
11. $\frac{24}{36}, \frac{4}{36}$, and $\frac{9}{36}$ 12. $\frac{80}{100}, \frac{5}{100}$, and $\frac{24}{100}$

EXERCISE 4B (PAGE 22)

1. $\frac{4}{8}$ and $\frac{1}{8}$ 2. $\frac{6}{10}$ and $\frac{3}{10}$ 3. $\frac{2}{24}$ and $\frac{5}{24}$
4. $\frac{8}{22}$ and $\frac{5}{22}$ 5. $\frac{5}{9}$ and $\frac{3}{9}$ 6. $\frac{6}{15}$ and $\frac{4}{15}$
7. $\frac{6}{16}, \frac{5}{16}$, and $\frac{4}{16}$ 8. $\frac{10}{20}, \frac{4}{20}$, and $\frac{3}{20}$
9. $\frac{6}{36}, \frac{4}{36}$, and $\frac{5}{36}$ 10. $\frac{4}{12}, \frac{2}{12}$, and $\frac{1}{12}$
11. $\frac{6}{30}, \frac{2}{30}$, and $\frac{11}{30}$ 12. $\frac{9}{27}, \frac{24}{27}$, and $\frac{11}{27}$

EXERCISE 4C (PAGE 23)

1. $\frac{15}{60}$ and $\frac{16}{60}$ 2. $\frac{2}{6}$ and $\frac{3}{6}$ 3. $\frac{9}{12}$ and $\frac{4}{12}$
4. $\frac{12}{21}$ and $\frac{14}{21}$ 5. $\frac{16}{20}$ and $\frac{15}{20}$ 6. $\frac{5}{10}$ and $\frac{8}{10}$
7. $\frac{21}{42}, \frac{14}{42}$, and $\frac{6}{42}$ 8. $\frac{28}{84}, \frac{21}{84}$, and $\frac{12}{84}$
9. $\frac{15}{60}, \frac{20}{60}$, and $\frac{12}{60}$ 10. $\frac{15}{30}, \frac{20}{30}$, and $\frac{6}{30}$
11. $\frac{40}{120}, \frac{48}{120}$, and $\frac{15}{120}$ 12. $\frac{45}{90}, \frac{54}{90}$, and $\frac{70}{90}$

EXERCISE 5A (PAGE 24)

1. $\frac{11}{15}$ 2. $\frac{5}{7}$ 3. $\frac{7}{12}$
4. $\frac{7}{8}$ 5. $\frac{9}{12}$ 6. $\frac{2}{3}$
7. $\frac{2}{3}$ 8. $\frac{1}{4}$ 9. $\frac{5}{6}$
10. $\frac{2}{3}$ 11. $\frac{6}{7}$ 12. $\frac{2}{11}$
13. $\frac{4}{9}$ 14. $\frac{1}{6}$ 15. $\frac{1}{3}$

Word Problem

Package B holds $\frac{7}{8}$ pound of rice, which is the same as $\frac{14}{16}$ pound. That is more than Package A holds.

$$\frac{7 \times 2}{8 \times 2} = \frac{14}{16}$$

EXERCISE 5B (PAGE 26)

1. $\frac{1}{4}, \frac{1}{3}, \frac{1}{2}$ 2. $\frac{1}{10}, \frac{1}{5}, \frac{1}{2}$ 3. $\frac{5}{16}, \frac{3}{8}, \frac{3}{4}$
4. $\frac{7}{24}, \frac{2}{3}, \frac{3}{4}$ 5. $\frac{1}{2}, \frac{2}{3}, \frac{5}{6}, \frac{7}{8}$ 6. $\frac{1}{4}, \frac{5}{12}, \frac{7}{12}, \frac{2}{3}$
7. $\frac{1}{8}, \frac{1}{6}, \frac{1}{4}$ 8. $\frac{1}{3}, \frac{7}{15}, \frac{4}{5}$ 9. $\frac{2}{5}, \frac{7}{15}, \frac{5}{9}$
10. $\frac{5}{16}, \frac{3}{8}, \frac{7}{12}, \frac{2}{3}$ 11. $\frac{5}{9}, \frac{2}{3}, \frac{3}{4}, \frac{5}{6}$ 12. $\frac{5}{14}, \frac{3}{7}, \frac{1}{2}, \frac{6}{7}$

Word Problem

From smallest to largest, the coins are **Coin D, Coin C, Coin A,** and **Coin B**.

Coin A: $\frac{7 \times 2}{10 \times 2} = \frac{14}{20}$

Coin B: $\frac{3 \times 5}{4 \times 5} = \frac{15}{20}$

Coin C: $\frac{3 \times 4}{5 \times 4} = \frac{12}{20}$

Coin D: $\frac{1 \times 10}{2 \times 10} = \frac{10}{20}$

EXERCISE 6A (PAGE 27)

PART A

1. **proper**   11. **improper**
2. **improper**   12. **improper**
3. **improper**   13. **improper**
4. **proper**   14. **improper**
5. **improper**   15. **improper**
6. **proper**   16. **proper**
7. **improper**   17. **improper**
8. **proper**   18. **improper**
9. **improper**   19. **proper**
10. **proper**   20. **improper**

Part B

1. $\frac{8}{5}$ 2. $\frac{5}{3}$ 3. $\frac{12}{7}$
4. $\frac{10}{6}$ 5. $\frac{9}{4}$ 6. $\frac{9}{3}$

Word Problem

The family ate $\frac{11}{8}$ tins of muffins. (The tins each have 8 parts, or muffins. Eleven parts are unshaded, so the improper fraction $\frac{11}{8}$ stands for the unshaded parts of the tins.)

EXERCISE 6B (PAGE 30)

1. $1\frac{1}{2}$ 2. $1\frac{1}{3}$ 3. $1\frac{2}{5}$
4. $2\frac{5}{9}$ 5. $3\frac{5}{8}$ 6. $4\frac{3}{8}$

Word Problem

The family ate $1\frac{5}{8}$ tins of muffins. (One whole tin and 5 parts of another are unshaded, so the mixed number $1\frac{5}{8}$ stands for the unshaded portion of the tins.)

EXERCISE 6C (PAGE 31)

1. $1\frac{1}{2}$ 2. $2\frac{1}{3}$ 3. 1 4. $1\frac{1}{3}$
5. $1\frac{3}{4}$ 6. $3\frac{1}{2}$ 7. $1\frac{3}{7}$ 8. 9
9. $1\frac{3}{4}$ 10. $1\frac{7}{8}$ 11. $1\frac{1}{4}$ 12. 2
13. $2\frac{4}{5}$ 14. $1\frac{3}{10}$ 15. $1\frac{3}{8}$ 16. 4
17. $1\frac{1}{6}$ 18. 3 19. 2 20. 1

Word Problem

Lena and her guests ate $2\frac{1}{2}$ pizzas.

$$\frac{20}{8} = 2\frac{4}{8} = 2\frac{1}{2}$$

EXERCISE 6D (PAGE 32)

1. $\frac{5}{2}$ 2. $\frac{34}{5}$ 3. $\frac{10}{3}$ 4. $\frac{17}{4}$
5. $\frac{25}{2}$ 6. $\frac{37}{9}$ 7. $\frac{29}{5}$ 8. $\frac{32}{3}$
9. $\frac{61}{7}$ 10. $\frac{53}{5}$ 11. $\frac{41}{10}$ 12. $\frac{16}{5}$
13. $\frac{62}{11}$ 14. $\frac{69}{16}$ 15. $\frac{73}{8}$ 16. $\frac{35}{6}$
17. $\frac{31}{4}$ 18. $\frac{29}{5}$ 19. $\frac{5}{3}$ 20. $\frac{59}{6}$

Word Problem

Archie walked $\frac{19}{8}$ miles.

To rewrite $2\frac{3}{8}$ as an improper fraction, multiply: $2 \times 8 = 16$.
Add the numerator to the product: $16 + 3 = 19$.
Write the sum over the denominator: $\frac{19}{8}$.

EXERCISE 7 (PAGE 35)

1. $\frac{3}{4}$ 2. $\frac{4}{5}$ 3. $\frac{9}{10}$ 4. $\frac{2}{3}$
5. $\frac{1}{2}$ 6. $\frac{5}{7}$ 7. $\frac{7}{9}$ 8. $\frac{4}{5}$
9. $1\frac{2}{3}$ 10. $\frac{1}{5}$ 11. $1\frac{1}{4}$ 12. 1
13. $\frac{3}{4}$ 14. $\frac{3}{5}$ 15. $\frac{5}{6}$ 16. 1
17. $1\frac{7}{9}$ 18. 1 19. $\frac{13}{16}$ 20. $1\frac{2}{9}$
21. $\frac{2}{3}$ 22. $1\frac{4}{5}$ 23. $1\frac{1}{12}$ 24. $\frac{14}{15}$

Word Problem

Juan spent $\frac{1}{2}$ of his workday filing and planning.

$$\frac{1}{8} + \frac{3}{8} = \frac{4 \div 4}{8 \div 4} = \frac{1}{2}$$

EXERCISE 8A (PAGE 37)

1. $7\frac{2}{3}$ 2. $11\frac{4}{5}$ 3. $7\frac{5}{6}$
4. $14\frac{1}{2}$ 5. $8\frac{6}{7}$ 6. $9\frac{1}{2}$
7. $3\frac{4}{9}$ 8. $8\frac{10}{11}$ 9. $15\frac{1}{2}$
10. $15\frac{3}{5}$ 11. $12\frac{2}{3}$ 12. $18\frac{10}{11}$
13. $14\frac{1}{2}$ 14. $25\frac{3}{4}$ 15. $43\frac{9}{19}$

Word Problem

The total length of board Shirley had left was **$15\frac{3}{4}$ inches.**

$$11\frac{3}{8} + 4\frac{3}{8} = 15\frac{6}{8}$$

$$15\frac{6 \div 2}{8 \div 2} = 15\frac{3}{4}$$

EXERCISE 8B (PAGE 38)

1. **8** 2. **12** 3. **17**
4. **11** 5. **9** 6. **14**
7. **19** 8. **64** 9. **21**
10. **34** 11. **16** 12. **14**
13. **15** 14. **19** 15. **67**

Word Problem

Susan spent **8 hours** in all watching television on Monday and Tuesday.

$$3\frac{1}{2} + 4\frac{1}{2} = 7\frac{2}{2}$$

Since $\frac{2}{2} = 1$, $7 + 1 = 8$.

EXERCISE 8C (PAGE 39)

1. $11\frac{2}{5}$ 2. $41\frac{1}{2}$ 3. $26\frac{1}{3}$
4. $12\frac{2}{5}$ 5. $24\frac{1}{3}$ 6. $7\frac{1}{11}$
7. $14\frac{1}{2}$ 8. $38\frac{3}{5}$ 9. $14\frac{4}{7}$
10. $5\frac{2}{3}$ 11. $19\frac{1}{3}$ 12. $15\frac{5}{16}$
13. $12\frac{3}{4}$ 14. $16\frac{8}{13}$ 15. $50\frac{1}{25}$

Word Problem

Manny worked $11\frac{3}{4}$ **hours** in all at his second job.

$$2\frac{1}{4} + 5\frac{3}{4} + 3\frac{3}{4} = 10\frac{7}{4}$$

Since $\frac{7}{4} = 1\frac{3}{4}$, $10 + 1\frac{3}{4} = 11\frac{3}{4}$.

EXERCISE 9 (PAGE 41)

1. $\frac{5}{8}$ 2. $\frac{9}{10}$ 3. $\frac{7}{24}$
4. $\frac{13}{22}$ 5. $\frac{8}{9}$ 6. $\frac{2}{3}$
7. $1\frac{1}{8}$ 8. $1\frac{1}{4}$ 9. $1\frac{5}{12}$
10. $1\frac{1}{5}$ 11. $\frac{31}{60}$ 12. $\frac{5}{6}$
13. $1\frac{1}{12}$ 14. $1\frac{5}{21}$ 15. $1\frac{11}{20}$
16. $1\frac{3}{10}$ 17. $1\frac{5}{24}$ 18. $1\frac{9}{20}$
19. $\frac{9}{40}$ 20. $\frac{43}{60}$ 21. $1\frac{1}{12}$
22. $1\frac{1}{10}$ 23. $\frac{15}{16}$ 24. $\frac{17}{20}$
25. $\frac{5}{12}$ 26. $\frac{7}{12}$ 27. $\frac{19}{30}$
28. $1\frac{1}{8}$ 29. 1 30. 1
31. $1\frac{7}{32}$ 32. $\frac{5}{6}$ 33. $\frac{41}{42}$
34. $\frac{61}{84}$ 35. $\frac{47}{60}$ 36. $1\frac{11}{30}$
37. $\frac{17}{18}$ 38. $1\frac{5}{12}$ 39. $1\frac{1}{12}$

Word Problem

Carrie ran $\frac{19}{20}$ **mile** in all.

$$\frac{1 \times 5}{4 \times 5} = \frac{5}{20}$$

$$\frac{1 \times 4}{5 \times 4} = \frac{4}{20}$$

$$\frac{1 \times 10}{2 \times 10} = \frac{10}{20}$$

$$\frac{5}{20} + \frac{4}{20} + \frac{10}{20} = \frac{19}{20}$$

EXERCISE 10 (PAGE 43)

1. $7\frac{7}{12}$ 2. $10\frac{5}{6}$ 3. $4\frac{3}{4}$
4. $9\frac{1}{6}$ 5. $4\frac{1}{4}$ 6. $6\frac{17}{20}$
7. $8\frac{1}{10}$ 8. $4\frac{31}{40}$ 9. $5\frac{11}{12}$
10. $8\frac{5}{8}$ 11. $6\frac{7}{16}$ 12. $7\frac{3}{10}$
13. $9\frac{1}{12}$ 14. $18\frac{19}{21}$ 15. $11\frac{1}{20}$
16. $14\frac{5}{12}$ 17. $7\frac{1}{12}$ 18. $14\frac{4}{5}$
19. $20\frac{35}{36}$ 20. $34\frac{7}{12}$

Word Problem

Altogether Carrie ran $6\frac{14}{15}$ **miles** that day.

$$2\frac{3 \times 5}{5 \times 3} = 2\frac{9}{15}$$

$$4\frac{1 \times 5}{3 \times 5} = 4\frac{5}{15}$$

$$2\frac{9}{15} + 4\frac{5}{15} = 6\frac{14}{15}$$

EXERCISE 11A (PAGE 45)

1. $\frac{3}{5}$ 2. $\frac{11}{23}$ 3. $\frac{2}{5}$
4. $\frac{13}{24}$ 5. $\frac{2}{7}$ 6. $\frac{7}{16}$
7. $\frac{1}{5}$ 8. $\frac{8}{13}$ 9. $\frac{5}{12}$
10. $\frac{7}{20}$ 11. $\frac{2}{3}$ 12. $\frac{2}{15}$
13. $\frac{5}{9}$ 14. $\frac{3}{11}$ 15. $\frac{2}{7}$
16. $\frac{1}{10}$ 17. $\frac{2}{5}$ 18. $\frac{5}{7}$
19. $\frac{7}{64}$ 20. $\frac{71}{100}$ 21. $\frac{3}{4}$
22. $\frac{1}{8}$ 23. $\frac{1}{5}$ 24. $\frac{1}{4}$
25. $\frac{2}{3}$ 26. $\frac{1}{3}$ 27. $\frac{2}{5}$
28. $\frac{1}{2}$ 29. $\frac{3}{4}$ 30. $\frac{5}{9}$
31. $\frac{1}{6}$ 32. $\frac{1}{3}$ 33. $\frac{5}{6}$
34. $\frac{1}{4}$ 35. $\frac{1}{18}$ 36. $\frac{4}{15}$
37. $\frac{1}{2}$ 38. $\frac{1}{4}$ 39. $\frac{2}{3}$

Word Problem

The pipe John had left was $\frac{1}{2}$ **yard** long.

$$\frac{7}{8} - \frac{3}{8} = \frac{4}{8}$$

$$\frac{4 \div 4}{8 \div 4} = \frac{1}{2}$$

EXERCISE 11B (PAGE 47)

1. $8\frac{2}{3}$ 2. $12\frac{2}{5}$ 3. $3\frac{2}{7}$
4. $20\frac{3}{11}$ 5. $1\frac{1}{3}$ 6. $6\frac{3}{5}$
7. $4\frac{1}{4}$ 8. $6\frac{1}{9}$ 9. $85\frac{1}{4}$
10. $129\frac{1}{2}$ 11. $5\frac{1}{2}$ 12. $7\frac{1}{3}$
13. $24\frac{3}{5}$ 14. $13\frac{2}{3}$ 15. $3\frac{1}{16}$
16. $15\frac{1}{3}$ 17. $5\frac{3}{5}$ 18. $7\frac{1}{3}$
19. $9\frac{1}{5}$ 20. $6\frac{1}{2}$

Word Problem

The packing materials weighed $3\frac{3}{4}$ **pounds.**

$$12\frac{15}{16} - 9\frac{3}{16} = 3\frac{12}{16}$$

$$3\frac{12 \div 4}{16 \div 4} = 3\frac{3}{4}$$

EXERCISE 12A (PAGE 48)

1. $\frac{5}{12}$ 2. $\frac{1}{15}$ 3. $\frac{1}{6}$
4. $\frac{1}{18}$ 5. $\frac{7}{24}$ 6. $\frac{3}{10}$
7. $\frac{11}{18}$ 8. $\frac{11}{20}$ 9. $\frac{3}{10}$
10. $\frac{37}{72}$ 11. $\frac{1}{2}$ 12. $\frac{1}{6}$
13. $\frac{3}{20}$ 14. $\frac{8}{15}$ 15. $\frac{11}{24}$
16. $\frac{9}{20}$ 17. $\frac{11}{60}$ 18. $\frac{13}{24}$
19. $\frac{3}{16}$ 20. $\frac{1}{2}$

Word Problem

Carlos walked $\frac{1}{2}$ **mile** a day after he hurt his ankle.

$$\frac{3 \times 2}{5 \times 2} = \frac{6}{10}$$

$$\frac{6}{10} - \frac{1}{10} = \frac{5}{10} = \frac{1}{2}$$

EXERCISE 12B (PAGE 50)

1. $3\frac{3}{8}$ 2. $4\frac{2}{15}$ 3. $7\frac{2}{21}$
4. $7\frac{1}{20}$ 5. $1\frac{7}{16}$ 6. $5\frac{1}{2}$
7. $5\frac{7}{12}$ 8. $26\frac{5}{9}$ 9. $13\frac{1}{3}$
10. $19\frac{3}{8}$ 11. $7\frac{5}{12}$ 12. $56\frac{1}{6}$
13. $17\frac{5}{16}$ 14. $1\frac{1}{32}$ 15. $73\frac{11}{28}$
16. $9\frac{4}{15}$ 17. $7\frac{5}{24}$ 18. $69\frac{4}{33}$
19. $1\frac{7}{20}$ 20. $101\frac{2}{21}$

Word Problem

Mason had $1\frac{7}{12}$ **tons** of grain left after the first week.

$$2\frac{3 \times 3}{4 \times 3} = 2\frac{9}{12}$$

$$1\frac{1 \times 2}{6 \times 2} = 1\frac{2}{12}$$

$$2\frac{9}{12} - 1\frac{2}{12} = 1\frac{7}{12}$$

EXERCISE 13A (PAGE 51)

1. $\frac{1}{2}$ 2. $\frac{1}{3}$ 3. $\frac{3}{4}$ 4. $\frac{2}{5}$
5. $\frac{1}{6}$ 6. $\frac{4}{7}$ 7. $\frac{3}{8}$ 8. $\frac{2}{9}$
9. $\frac{7}{10}$ 10. $\frac{13}{16}$ 11. $\frac{10}{13}$ 12. $\frac{8}{21}$

Word Problem

After Dionne cut off the part she needed, a $\frac{3}{16}$-**inch** piece of the molding was left.

$$1 - \frac{13}{16} = \frac{3}{16}$$

EXERCISE 13B (PAGE 52)

1. $6\frac{1}{2}$ 2. $10\frac{2}{5}$ 3. $20\frac{7}{16}$ 4. $31\frac{4}{9}$
5. $11\frac{1}{3}$ 6. $5\frac{6}{11}$ 7. $8\frac{5}{12}$ 8. $12\frac{3}{7}$
9. $20\frac{5}{6}$ 10. $10\frac{7}{8}$ 11. $2\frac{1}{2}$ 12. $8\frac{1}{4}$
13. $6\frac{7}{16}$ 14. $13\frac{1}{6}$ 15. $26\frac{2}{3}$ 16. $62\frac{3}{8}$
17. $13\frac{2}{9}$ 18. $18\frac{2}{5}$ 19. $40\frac{1}{8}$ 20. $199\frac{4}{5}$

Word Problem

After he painted the shutters, Tony had $3\frac{1}{4}$ **gallons** of blue paint left.

$$4 - \frac{3}{4} = 3\frac{1}{4}$$

EXERCISE 13C (PAGE 54)

1. $3\frac{2}{5}$ 2. $1\frac{4}{9}$ 3. $\frac{1}{3}$
4. $11\frac{3}{4}$ 5. $4\frac{1}{2}$ 6. $2\frac{4}{5}$
7. $8\frac{1}{4}$ 8. $\frac{2}{3}$ 9. $1\frac{3}{4}$
10. $8\frac{7}{8}$ 11. $\frac{4}{7}$ 12. $\frac{1}{2}$

Word Problem

The piece of fabric that was left was $24\frac{1}{2}$ **inches** long.

Step 1	Step 2	Step 3 and 4
$35\frac{3}{8}$	$34\frac{8}{8} + \frac{3}{8}$	$34\frac{11}{8}$
$-10\frac{7}{8}$	$-10\frac{7}{8}$	$-10\frac{7}{8}$
		$24\frac{4}{8} = 24\frac{1}{2}$

EXERCISE 13D (PAGE 55)

1. $3\frac{13}{15}$ 2. $1\frac{13}{18}$ 3. $8\frac{2}{3}$
4. $8\frac{5}{18}$ 5. $3\frac{5}{12}$ 6. $6\frac{11}{15}$
7. $11\frac{3}{4}$ 8. $5\frac{19}{20}$ 9. $30\frac{5}{6}$
10. $7\frac{1}{2}$ 11. $6\frac{7}{18}$ 12. $3\frac{3}{4}$
13. $13\frac{5}{6}$ 14. $46\frac{9}{10}$ 15. $38\frac{5}{8}$
16. $12\frac{3}{4}$ 17. $38\frac{15}{16}$ 18. $12\frac{9}{10}$
19. $9\frac{9}{20}$ 20. $47\frac{17}{24}$

Word Problem

Sam hauled $4\frac{3}{4}$ **truckloads** more on Tuesday than on Wednesday.

Step 1	Step 2 and 3
$12\frac{1}{4} = 12\frac{1}{4}$	$\overset{11\frac{4}{4}}{\cancel{12}} + \frac{1}{4} = 11\frac{5}{4}$
$7\frac{1 \times 2}{2 \times 2} = 7\frac{2}{4}$	$-7\frac{2}{4} = 7\frac{2}{4}$
	$4\frac{3}{4}$

MIXED PRACTICE 1 (page 56)

1. $\frac{5}{7}$ 2. $\frac{3}{5}$ 3. $\frac{2}{3}$
4. $\frac{3}{4}$ 5. 1 6. $\frac{1}{5}$
7. $3\frac{2}{9}$ 8. $5\frac{3}{5}$ 9. $8\frac{2}{3}$
10. $7\frac{3}{4}$ 11. $15\frac{1}{2}$ 12. $7\frac{1}{3}$
13. 8 14. 12 15. 19
16. $11\frac{2}{5}$ 17. $14\frac{1}{2}$ 18. $12\frac{3}{4}$
19. $\frac{8}{9}$ 20. $\frac{1}{15}$ 21. $\frac{19}{30}$
22. $\frac{5}{6}$ 23. $\frac{5}{6}$ 24. $\frac{47}{60}$
25. $1\frac{11}{30}$ 26. $\frac{8}{15}$ 27. $\frac{11}{60}$
28. $\frac{1}{2}$ 29. $\frac{17}{18}$ 30. $1\frac{27}{40}$
31. $4\frac{1}{4}$ 32. $5\frac{11}{12}$ 33. $3\frac{3}{8}$

34. $5\frac{7}{12}$ 35. $14\frac{4}{5}$ 36. $13\frac{1}{3}$
37. $\frac{13}{16}$ 38. $2\frac{3}{7}$ 39. $499\frac{1}{4}$
40. $3\frac{2}{5}$ 41. $\frac{1}{3}$ 42. $1\frac{3}{4}$
43. $3\frac{13}{15}$ 44. $6\frac{7}{18}$ 45. $12\frac{3}{4}$

EXERCISE 14 (PAGE 59)

1. $\frac{1}{16}$ 2. $\frac{2}{15}$ 3. $\frac{3}{32}$
4. $\frac{2}{21}$ 5. $\frac{4}{21}$ 6. $\frac{1}{4}$
7. $\frac{9}{80}$ 8. $\frac{1}{6}$ 9. $\frac{4}{9}$
10. $\frac{1}{10}$ 11. $\frac{3}{64}$ 12. $\frac{1}{16}$
13. $\frac{1}{27}$ 14. $\frac{4}{15}$ 15. $\frac{1}{30}$
16. $\frac{5}{18}$ 17. $\frac{1}{24}$ 18. $\frac{2}{15}$
19. $\frac{9}{16}$ 20. $\frac{5}{24}$

Word Problem

Carrie used $\frac{2}{9}$ **gallon** of paint.

$$\frac{2}{3} \times \frac{1}{3} = \frac{2}{9}$$

EXERCISE 15A (PAGE 60)

1. $\frac{1}{8}$ 2. $\frac{3}{10}$ 3. $\frac{4}{25}$
4. $\frac{4}{7}$ 5. $\frac{1}{9}$ 6. $\frac{2}{9}$
7. $\frac{5}{8}$ 8. $\frac{15}{22}$ 9. $\frac{3}{14}$
10. $\frac{3}{8}$ 11. $\frac{10}{27}$ 12. $\frac{5}{12}$
13. $\frac{7}{12}$ 14. $\frac{4}{9}$ 15. $\frac{1}{10}$

Word Problem

Tony was allowed to keep $\frac{3}{32}$ **ounce** of gold.

$$\frac{\overset{3}{\cancel{15}}}{16} \times \frac{1}{\underset{2}{\cancel{10}}} = \frac{3}{32}$$

EXERCISE 15B (PAGE 62)

1. $\frac{3}{4}$ 2. $\frac{3}{10}$ 3. $\frac{3}{7}$
4. $\frac{3}{8}$ 5. $\frac{1}{6}$ 6. $\frac{2}{3}$
7. $\frac{1}{4}$ 8. $\frac{1}{12}$ 9. $\frac{1}{4}$
10. $\frac{1}{2}$ 11. $\frac{1}{6}$ 12. $\frac{1}{2}$

Word Problem

Tony kepy $\frac{1}{4}$ **ounce** of gold.

$$\frac{\overset{1}{\cancel{4}}}{9} \times \frac{\overset{1}{\cancel{9}}}{\underset{4}{\cancel{16}}} = \frac{1}{4}$$

EXERCISE 16 (PAGE 63)

1. $1\frac{1}{4}$ 2. $\frac{1}{3}$ 3. $3\frac{1}{3}$
4. $2\frac{2}{3}$ 5. $6\frac{1}{8}$ 6. $\frac{8}{9}$
7. $4\frac{2}{3}$ 8. $1\frac{5}{7}$ 9. $1\frac{1}{8}$
10. $2\frac{2}{5}$ 11. $4\frac{4}{5}$ 12. $4\frac{1}{12}$
13. 9 14. 3 15. $3\frac{2}{3}$

Word Problem

Teresa would charge **$18** if it took her $\frac{3}{4}$ hour to fix a carburetor.

$$\frac{\overset{6}{\cancel{24}}}{1} \times \frac{3}{\underset{1}{\cancel{4}}} = \frac{18}{1} = 18$$

EXERCISE 17 (PAGE 64)

1. $4\frac{1}{8}$ 2. $3\frac{3}{8}$ 3. $1\frac{13}{15}$
4. $1\frac{13}{27}$ 5. $7\frac{1}{5}$ 6. $1\frac{1}{4}$
7. $13\frac{1}{8}$ 8. $2\frac{1}{4}$ 9. $\frac{1}{2}$
10. $\frac{2}{3}$ 11. 4 12. $3\frac{7}{20}$
13. $1\frac{1}{9}$ 14. $\frac{23}{28}$ 15. $\frac{2}{3}$

Word Problem

The board Nancy needed was $2\frac{1}{3}$ **feet** long.

$$\frac{7}{\underset{1}{\cancel{4}}} \times \frac{\overset{1}{\cancel{4}}}{3} = \frac{7}{3} = 2\frac{1}{3}$$

EXERCISE 18A (PAGE 67)

1. 2 2. 4 3. $1\frac{1}{6}$
4. $1\frac{1}{3}$ 5. $1\frac{1}{8}$ 6. $1\frac{3}{7}$
7. $\frac{2}{3}$ 8. $1\frac{1}{2}$ 9. $1\frac{1}{2}$
10. $\frac{5}{6}$ 11. $2\frac{1}{4}$ 12. 1
13. $1\frac{2}{3}$ 14. $\frac{3}{5}$ 15. $2\frac{3}{16}$
16. $1\frac{1}{3}$ 17. 1 18. $1\frac{1}{5}$
19. $1\frac{1}{2}$ 20. $\frac{2}{3}$

Word Problem

Denise can divide the right-of-way into **6 strips.**

Step 1	Step 2	Steps 3 and 4
$\frac{3}{4} \div \frac{1}{8} =$	$\frac{3}{4} \times \frac{8}{1} =$	$\frac{3}{\underset{1}{\cancel{4}}} \times \frac{\overset{2}{\cancel{8}}}{1} = \frac{6}{1} = 6$

EXERCISE 18B (PAGE 68)

1. **4** 2. **$7\frac{2}{9}$** 3. **10**
4. **6** 5. **$3\frac{8}{9}$** 6. **$10\frac{2}{7}$**
7. **9** 8. **$6\frac{1}{2}$** 9. **2**
10. **$1\frac{5}{8}$** 11. **7** 12. **5**
13. **$4\frac{1}{2}$** 14. **18** 15. **$3\frac{1}{2}$**
16. **24** 17. **$6\frac{1}{2}$** 18. **12**
19. **$1\frac{1}{2}$** 20. **$13\frac{1}{2}$**

Word Problem

Prasong will have **10 shelves**.

Step 1	Step 2	Step 3
$2\frac{1}{2} \div \frac{1}{4} =$	$\frac{5}{2} \div \frac{1}{4} =$	$\frac{5}{2} \times \frac{4}{1} =$

Step 4	Step 5
$\frac{5}{\cancel{2}_1} \times \frac{\cancel{4}^2}{1} = \frac{10}{1}$	$\frac{10}{1} = 10$

EXERCISE 19A (PAGE 70)

1. $\frac{2}{9}$ 2. $\frac{4}{15}$ 3. $\frac{1}{12}$
4. $\frac{7}{16}$ 5. $\frac{1}{4}$ 6. $\frac{3}{10}$
7. $\frac{1}{10}$ 8. $\frac{1}{15}$ 9. $\frac{1}{14}$
10. $\frac{1}{15}$ 11. $\frac{3}{28}$ 12. $\frac{1}{25}$
13. $\frac{1}{6}$ 14. $\frac{3}{32}$ 15. $\frac{7}{64}$
16. $\frac{1}{9}$ 17. $\frac{1}{18}$ 18. $\frac{15}{112}$
19. $\frac{2}{27}$ 20. $\frac{1}{10}$

Word Problem

Each person will get $\frac{1}{8}$ **pound** of chocolate.

Step 1	Step 2	Step 3	Step 4
$\frac{3}{4} \div 6 =$	$\frac{3}{4} \div \frac{6}{1} =$	$\frac{3}{4} \times \frac{1}{6} =$	$\frac{\cancel{3}^1}{4} \times \frac{1}{\cancel{6}_2} = \frac{1}{8}$

EXERCISE 19B (PAGE 71)

1. $\frac{5}{8}$ 2. $\frac{1}{16}$ 3. $\frac{1}{3}$
4. $\frac{1}{5}$ 5. $\frac{3}{4}$ 6. $\frac{1}{16}$
7. $\frac{1}{6}$ 8. $\frac{1}{10}$ 9. $\frac{3}{16}$
10. $\frac{3}{32}$ 11. $\frac{3}{7}$ 12. $\frac{3}{8}$

Word Problem

Each child would get $\frac{1}{4}$ **pie**.

Step 1	Step 2	Step 3	Step 4
$1\frac{1}{4} \div 5 =$	$\frac{5}{4} \div \frac{5}{1} =$	$\frac{5}{4} \times \frac{1}{5} =$	$\frac{\cancel{5}^1}{4} \times \frac{1}{\cancel{5}_1} = \frac{1}{4}$

EXERCISE 20 (PAGE 72)

1. **$1\frac{2}{5}$** 2. **$1\frac{3}{11}$** 3. **1**
4. $\frac{9}{40}$ 5. **$1\frac{1}{4}$** 6. $\frac{4}{9}$
7. $\frac{7}{20}$ 8. **1** 9. **$1\frac{4}{5}$**
10. **5** 11. $\frac{3}{4}$ 12. $\frac{4}{41}$
13. $\frac{5}{8}$ 14. $\frac{4}{19}$ 15. **$1\frac{1}{25}$**
16. **$1\frac{11}{20}$** 17. **$3\frac{4}{7}$** 18. **$2\frac{4}{9}$**
19. **$1\frac{1}{5}$** 20. **$6\frac{3}{4}$**

Word Problem

Luz can make **4 dresses**.

Step 1	Step 2
$13\frac{1}{3} \div 3\frac{1}{3} =$	$\frac{40}{3} \div \frac{10}{3} =$

Step 3	Steps 4 and 5
$\frac{40}{3} \times \frac{3}{10} =$	$\frac{\cancel{40}^4}{\cancel{3}_1} \times \frac{\cancel{3}^1}{\cancel{10}_1} = \frac{4}{1} = 4$

MIXED PRACTICE 2 (page 74)

1. $\frac{3}{8}$ 2. **$1\frac{1}{5}$** 3. $\frac{2}{15}$
4. **2** 5. $\frac{4}{21}$ 6. **$1\frac{3}{7}$**
7. **$2\frac{1}{4}$** 8. $\frac{4}{9}$ 9. $\frac{1}{30}$
10. **$1\frac{1}{3}$** 11. **$3\frac{3}{5}$** 12. $\frac{2}{21}$
13. **4** 14. $\frac{3}{10}$ 15. $\frac{4}{25}$
16. **$10\frac{2}{7}$** 17. **7** 18. $\frac{3}{14}$
19. $\frac{10}{27}$ 20. **24** 21. $\frac{1}{6}$
22. $\frac{1}{6}$ 23. $\frac{2}{9}$ 24. $\frac{3}{4}$
25. $\frac{3}{10}$ 26. $\frac{3}{28}$ 27. $\frac{3}{4}$
28. $\frac{1}{9}$ 29. $\frac{1}{4}$ 30. $\frac{1}{2}$
31. **$6\frac{2}{5}$** 32. $\frac{6}{25}$ 33. $\frac{5}{8}$
34. $\frac{1}{5}$ 35. **$1\frac{1}{4}$** 36. **$2\frac{2}{3}$**
37. **$4\frac{2}{3}$** 38. $\frac{3}{16}$ 39. $\frac{3}{7}$

40. $4\frac{1}{12}$ 41. $1\frac{2}{5}$ 42. $1\frac{7}{15}$
43. $1\frac{7}{9}$ 44. $4\frac{1}{8}$ 45. $\frac{4}{9}$
46. $1\frac{13}{27}$ 47. $1\frac{1}{25}$ 48. $9\frac{3}{8}$
49. $1\frac{1}{9}$ 50. $6\frac{3}{4}$

FRACTION SKILLS REVIEW (page 75)

Part A
1. $\frac{2}{3}$ 2. $\frac{5}{7}$ 3. $\frac{5}{8}$ 4. $\frac{1}{4}$ 5. $\frac{2}{3}$

Part B
1. $\frac{10}{12}$ 2. $\frac{5}{20}$ 3. $\frac{21}{56}$ 4. $\frac{8}{72}$ 5. $\frac{16}{24}$

Part C
1. $\frac{7}{10}$ 2. $\frac{4}{7}$ 3. $\frac{5}{8}$ 4. $\frac{17}{18}$ 5. $\frac{7}{8}$

Part D
1. $\frac{3}{7}, \frac{2}{5}, \frac{1}{3}$ 2. $\frac{3}{4}, \frac{5}{8}, \frac{5}{16}$ 3. $\frac{3}{4}, \frac{5}{7}, \frac{2}{3}$
4. $\frac{4}{9}, \frac{3}{8}, \frac{2}{7}$ 5. $\frac{2}{5}, \frac{3}{8}, \frac{1}{3}$

Part E
1. $1\frac{1}{2}$ 2. $1\frac{3}{7}$ 3. 1 4. $1\frac{1}{2}$ 5. 3

Part F
1. $\frac{5}{2}$ 2. $\frac{13}{4}$ 3. $\frac{14}{5}$ 4. $\frac{29}{11}$ 5. $\frac{29}{9}$

Part G
1. $\frac{5}{8}$ 2. $\frac{1}{4}$ 3. $\frac{2}{15}$
4. $\frac{1}{2}$ 5. 2 6. $1\frac{1}{3}$
7. $\frac{2}{21}$ 8. 1 9. $\frac{2}{25}$
10. $8\frac{5}{9}$ 11. 10 12. $32\frac{2}{23}$
13. $3\frac{1}{3}$ 14. $\frac{2}{7}$ 15. $\frac{3}{5}$
16. $3\frac{3}{4}$ 17. 6 18. $\frac{5}{9}$
19. $\frac{17}{18}$ 20. $\frac{7}{12}$ 21. $\frac{7}{8}$
22. $2\frac{5}{8}$ 23. $1\frac{1}{2}$ 24. $11\frac{1}{15}$
25. $\frac{13}{18}$ 26. $\frac{1}{20}$ 27. $8\frac{1}{32}$
28. $\frac{1}{5}$ 29. $7\frac{2}{3}$ 30. 28
31. $50\frac{11}{12}$ 32. 15 33. $1\frac{1}{2}$

EXERCISE 21A (PAGE 78)

1. $1\frac{1}{4}$ 2. $2\frac{3}{4}$
3. $2\frac{1}{4}$ 4. $3\frac{3}{8}$
5. $\frac{3}{4}$ 6. $2\frac{1}{2}$

7. $\frac{3}{4}$ 8. $\frac{1}{4}$
9. $1\frac{1}{2}$ 10. $18\frac{1}{2}$

Word Problem
Sylvia needs $1\frac{2}{3}$ **yards** of material.

$$\frac{5}{1} \div \frac{3}{1}$$

$$\frac{5}{1} \times \frac{1}{3} = \frac{5}{3} = 1\frac{2}{3}$$

EXERCISE 21B (PAGE 79)

1. $4\frac{1}{2}$ 2. **90**
3. **76**   4. **33**
5. **64**   6. **18**
7. **18,480**   8. **51**
9. **18**   10. **11,220**

Word Problem
Jamal needs **5 cups** of water.

$$2\frac{1}{2} \times 2 =$$

$$\frac{5}{2} \times \frac{2}{1} = \frac{5}{1} = 5$$

EXERCISE 21C (PAGE 81)

1. **72**   2. $3\frac{3}{4}$
3. **720**   4. $2\frac{1}{3}$
5. $\frac{3}{4}$ 6. $\frac{1}{10}$
7. $5866\frac{2}{3}$ 8. **64**
9. **20**   10. $6\frac{1}{4}$

Word Problem
Sylvia needs $1\frac{5}{6}$ **yd** of cloth.

Step 1 **Step 2** **Step 3**

1 ft = 12 in. $\frac{66}{1} \div \frac{12}{1} =$ $\frac{66}{1} \times \frac{1}{12} = \frac{33}{6} = 5\frac{1}{2}$

Step 4 **Step 5** **Step 6**

1 yd = 3 ft $5\frac{1}{2} \div 3$ $\frac{11}{2} \times \frac{1}{3} = \frac{11}{6} = 1\frac{5}{6}$

$\frac{11}{2} \div \frac{3}{1}$

EXERCISE 22A (PAGE 82)

1. $11\frac{3}{4}$ in.
2. $24\frac{7}{12}$ yd
3. 9 ft
4. $12\frac{3}{8}$ in.
5. $72\frac{1}{2}$ mi
6. $5\frac{1}{3}$ in.
7. $16\frac{3}{4}$ ft
8. $23\frac{1}{4}$ yd
9. $16\frac{2}{3}$ ft
10. 15 mi
11. $24\frac{3}{4}$ ft
12. 30 yd
13. $17\frac{3}{5}$ ft
14. $24\frac{2}{3}$ in.
15. $85\frac{1}{3}$ yd

Word Problem

To enclose the garden, $15\frac{1}{2}$ **yards** of fencing are needed.

$$3\frac{1}{2} = 3\frac{2}{4}$$
$$+ 4\frac{1}{4} = 4\frac{1}{4}$$
$$+ 3\frac{1}{2} = 3\frac{2}{4}$$
$$+ 4\frac{1}{4} = 4\frac{1}{4}$$
$$14\frac{6}{4} = 15\frac{2}{4} = 15\frac{1}{2}$$

EXERCISE 22B (PAGE 85)

1. $5\frac{1}{2}$ ft
2. $2\frac{3}{4}$ in.
3. $6\frac{3}{5}$ mi
4. $3\frac{7}{16}$ in.
5. 4 yd
6. 2 mi

Word Problem

The circumference of the pool is $13\frac{3}{4}$ **feet**.

Circumference $= \frac{22}{7} \times$ diameter

Circumference $= \frac{22}{7} \times 4\frac{3}{8}$ ft

Circumference $= \frac{\overset{11}{\cancel{22}}}{\cancel{7}} \times \frac{\overset{5}{\cancel{35}}}{\cancel{8}}$ ft $= \frac{55}{4} = 13\frac{3}{4}$ ft

EXERCISE 22C (PAGE 86)

Part A

1. $12\frac{1}{4}$ sq in.
2. 49 sq ft
3. $\frac{11}{56}$ sq ft
4. $3\frac{1}{16}$ sq ft
5. $65\frac{1}{3}$ sq ft
6. $\frac{7}{22}$ sq mi
7. $1\frac{13}{36}$ sq yd
8. $12\frac{24}{25}$ sq in.
9. $17\frac{1}{9}$ sq yd
10. $56\frac{1}{4}$ sq ft

Part B

1. 156 sq in.
2. $160\frac{4}{9}$ sq yd
3. $4\frac{13}{18}$ sq in.
4. $5\frac{1}{11}$ sq in.
5. $76\frac{9}{16}$ sq ft
6. $40\frac{3}{10}$ sq mi
7. $\frac{11}{350}$ sq mi
8. $14\frac{2}{3}$ sq in.
9. $26\frac{22}{49}$ sq in.
10. $77\frac{35}{48}$ sq yd

Word Problem

The area of the garden is **14 square yards**.

Area = length × width

Area $= 3\frac{1}{2}$ yd × 4 yd

Area $= \frac{7}{\cancel{2}} \times \frac{\overset{2}{\cancel{4}}}{1} = \frac{14}{1} = 14$ sq yd

EXERCISE 23A (PAGE 89)

1. .2
2. .5
3. .25
4. .75
5. .8
6. .833
7. .3
8. .167
9. .875
10. .222
11. .6
12. .083

Word Problem

The value of a share of the stock declines **$.375** if it goes down $\frac{3}{8}$.

$$\frac{3}{8} = 3 \div 8 = .375$$

EXERCISE 23B (PAGE 90)

1. $\frac{7}{20}$
2. $\frac{7}{1000}$
3. $\frac{9}{500}$
4. $\frac{3}{10}$
5. $\frac{33}{50}$
6. $\frac{3}{4}$
7. $\frac{2}{25}$
8. $\frac{3}{40}$
9. $\frac{1}{5}$
10. $\frac{4}{5}$
11. $\frac{1}{10,000}$
12. $\frac{5}{8}$

Word Problem

The diameter of Harry's drill bit is $\frac{1}{8}$ **inch**.

$$\frac{125 \div 125}{1000 \div 125} = \frac{1}{8}$$

EXERCISE 23C (PAGE 91)

1. Using decimals: Roger ate **.2 pound** of cheese.
   .25 × .8 lb = .2 lb
   Using fractions: Roger ate $\frac{1}{5}$ **pound** of cheese.

$$\frac{1}{\cancel{4}} \times \frac{\overset{2}{\cancel{8}}}{10} = \frac{2}{10} = \frac{1}{5}$$

2. Karin paid **$2.20** for the potatoes.
   2.75 × $.80 = $2.20

3. Using decimals:

   It will take Mariama **29.75 minutes** to jog 3.5 miles.
   $8.5 \times 3.5 = 29.75$ min

   Using fractions:

   It will take Mariama $29\frac{3}{4}$ **minutes** to jog $3\frac{1}{2}$ miles.
   $\frac{17}{2} \times \frac{7}{2} = \frac{119}{4} = 29\frac{3}{4}$

4. Luz paid **$9.45** for the cloth.
   $1.75 \times \$5.40 = \$9.45$

5. Using decimals:

   Beng Choo received **.9 ton** of feed the next day.
   $.25 \times 3.6 = .9$ T

   Using fractions:

   Beng Choo received $\frac{9}{10}$ **ton** of feed the next day.
   $\frac{1}{4} \times \frac{\overset{9}{\cancel{36}}}{10} = \frac{9}{10}$ T

EXERCISE 24A (PAGE 92)

1. **9**	2. **24**	3. **12**	4. **8**
5. **17**	6. **15**	7. **10**	8. **30**
9. **15**	10. **7**	11. **12**	12. **100**
13. **55**	14. **67**	15. **46**	

Word Problem

   Jack should estimate **6 hours** for this job.
   ($\frac{1}{4}$ of an hour is less than $\frac{1}{2}$, or $\frac{2}{4}$, of an hour.)

EXERCISE 24B (PAGE 93)

1. $14 + 8 =$ **22**
2. $6 \div 2 =$ **3**
3. $3 \times 5 =$ **15**
4. $25 - 15 =$ **10**
5. $10 \times 6 =$ **60**
6. $48 \div 8 =$ **6**
7. $30 + 70 =$ **100**
8. $37 - 17 =$ **20**
9. $9 \div 3 =$ **3**
10. $83 - 40 =$ **43**
11. $4 \times 25 =$ **100**
12. $20 + 20 =$ **40**
13. $39 \div 13 =$ **3**
14. $50 \times 20 =$ **1000**
15. $54 - 45 =$ **9**

Word Problem

   William's estimate was **10 bolts**.
   $18 - 8 = 10$

EXERCISE 25 (PAGE 97)

1. John harvested $3\frac{7}{8}$ **tons** of wheat altogether.
   $1\frac{1}{2} + 2\frac{3}{8} = 3\frac{7}{8}$

2. Emily can fertilize **4 rows**.
   $9 \div 2\frac{1}{4} = 4$

3. Earl needs **37 feet** of metal shelving.
   $9\frac{1}{4} \times 4 = 37$

4. Altogether, Andrea worked $6\frac{1}{4}$ **days**.
   $3\frac{3}{4} + 2\frac{1}{2} = 6\frac{1}{4}$

5. Fred needs **38 yards** of rope.
   $12 \times 3\frac{1}{6} = 38$

6. The blackfish weighed $5\frac{1}{4}$ **pounds**.
   $6 - \frac{3}{4} = 5\frac{1}{4}$

7. The computer saves Joan $1\frac{3}{4}$ **hours** on each weekly report.
   $3\frac{1}{2} - 1\frac{3}{4} = 1\frac{3}{4}$

8. Altogether the two neighbors bought $15\frac{3}{4}$ **yards** of screening material.
   $3\frac{1}{4} + 12\frac{1}{2} = 15\frac{3}{4}$

9. Kathy used $2\frac{3}{8}$ **gallons** of paint.
   $4\frac{3}{4} \times \frac{1}{2} = 2\frac{3}{8}$

10. Dimitri earns **$39**.
    $12 \times 3\frac{1}{4} = 39$

11. Each person will get $\frac{1}{8}$ **gallon** of ice cream.
    $\frac{1}{2} \div 4 = \frac{1}{8}$

12. Celia had **16 portions**.
    $2\frac{2}{3} \div \frac{1}{6} = 16$

13. It takes Juan $6\frac{1}{4}$ **hours** to stand and prime 5 table tops.
    $1\frac{1}{4} \times 5 = 6\frac{1}{4}$

14. Felix had $\frac{3}{4}$ **can** of paint thinner altogether.
    $\frac{1}{3} + \frac{1}{4} + \frac{1}{6} = \frac{3}{4}$

15. Sing has **30 feet** of bookshelves altogether.
    $9 \times 3\frac{1}{3} = 30$

16. Joanna needs another $\frac{7}{8}$ **pint** of beef stock.
    $1\frac{3}{4} - \frac{7}{8} = \frac{7}{8}$

17. Each hamburger weighed $\frac{1}{4}$ **pound.**
    $2\frac{1}{2} \div 10 = \frac{1}{4}$

18. Antonio had $4\frac{1}{2}$ **feet** of matting left over.
    $45 - 40\frac{1}{2} = 4\frac{1}{2}$

19. It would take the juicer $5\frac{1}{2}$ **minutes** to make $13\frac{3}{4}$ cups of orange juice.
    $13\frac{3}{4} \div 2\frac{1}{2} = 5\frac{1}{2}$

20. $2\frac{7}{12}$ dozen rolls of paper towels were not delivered.
    $15 - 12\frac{5}{12} = 2\frac{7}{12}$

EXERCISE 26 (PAGE 99)

Part A

1. (3)   2. (2)   3. (2)   4. (5)   5. (4)

Part B

1. (5) $3\frac{1}{2}$ hr × $7 = $24.50;
   $24.50 × 5 da = $122.50

2. (2) $6\frac{1}{4}$ yd × $12 = $75;
   $100 − $75 = $25

3. (1) $2\frac{1}{4}$ mi × $3\frac{1}{3}$ mi = $7\frac{1}{2}$ sq mi;
   $7\frac{1}{2}$ sq mi × 300 sacks = 2250 sacks

4. (4) $8\frac{1}{4}$ ft × 2 = $16\frac{1}{2}$ ft;
   $16\frac{1}{2}$ ft + $8\frac{1}{4}$ ft = $24\frac{3}{4}$ ft

5. (3) 120 yd + 39 yd + 120 yd + 39 yd
   = 318 yd;
   318 yd ÷ $\frac{3}{4}$ yd per step = 424 steps

Part C

1. Glen paid **$52** altogether.
   $39 × $\frac{2}{3}$ = $26;
   $26 × 2 lamps = $52

2. The cost of enough plastic to cover the pool is **$192.**
   $5\frac{1}{3}$ yd × 9 yd = 48 sq yd;
   48 sq yd × $4 = $192

3. Millen will use another **8 gallons** of fuel.
   210 mi ÷ $10\frac{1}{2}$ gal = 20 mi per gal;
   160 mi ÷ 20 mi per gal = 8 gal

4. Jim paid **$96** in all.
   $24 + $8 + $96 = $128;
   $128 × $\frac{3}{4}$ = $96

5. Mino needs **108 tiles** in all.
   12 ft × 16 ft = 192 sq ft;
   $1\frac{1}{3}$ ft × $1\frac{1}{3}$ ft = $\frac{16}{9}$ sq ft per tile;
   192 sq ft ÷ $\frac{16}{9}$ sq ft per tile = 108 tiles

FRACTIONS REVIEW (page 103)

1. $1\frac{1}{20}$ 2. $\frac{3}{8}$ 3. $\frac{1}{3}$
4. $\frac{2}{3}$ 5. $4\frac{2}{5}$ 6. $12\frac{3}{4}$
7. $\frac{5}{12}$ 8. $\frac{7}{12}$ 9. $10\frac{7}{20}$
10. $\frac{17}{18}$ 11. $\frac{4}{9}$ 12. $1\frac{1}{2}$
13. $2\frac{2}{3}$ 14. $\frac{1}{2}$ 15. $4\frac{11}{12}$
16. $2\frac{3}{4}$ 17. $3\frac{2}{3}$ 18. $\frac{1}{10}$
19. $\frac{2}{3}$ 20. $\frac{3}{8}$ 21. $\frac{14}{15}$
22. $1\frac{1}{5}$ 23. $3\frac{1}{2}$ 24. **7**
25. **2 in.**   26. $\frac{1}{6}$ **sq in.**   27. $4\frac{3}{4}$ **in.**
28. $1\frac{5}{16}$ **sq in.**   29. $\frac{4}{5}$ **mi**   30. $\frac{1}{25}$ **sq mi**
31. **11 ft**   32. $9\frac{5}{8}$ **sq ft**

33. Hector should use a $\frac{5}{8}$-**inch** drill bit.
    .625 in. = $\frac{625}{1000}$ in. = $\frac{5}{8}$ in.

34. Minerva can make **4 dresses.**
    $6\frac{1}{2} \div 1\frac{5}{8} = 4$

35. It takes Ali **7 minutes.**
    Using decimals:
    3.5 mi × .25 = .875 mi;
    8 min per mi × .875 mi = 7 min
    Using fractions:
    $3\frac{1}{2}$ mi × $\frac{1}{4}$ = $\frac{7}{8}$ mi;
    8 min per mi × $\frac{7}{8}$ mi = 7 min

GED PRACTICE 1 (page 105)

1. (4) $\frac{1}{4}$ gal + $1\frac{1}{2}$ gal = $1\frac{3}{4}$ gal
2. (2) $\frac{5}{8}$ C − $\frac{3}{8}$ C = $\frac{1}{4}$ C

3. **(3)** $6\frac{3}{4}$ lb $- 2\frac{1}{8}$ lb $= 4\frac{5}{8}$ lb
4. **(5)** $4\frac{1}{4}$ mi $- 1\frac{1}{2}$ mi $= 2\frac{3}{4}$ mi
5. **(1)** $\frac{4}{5}$ qt $\times \frac{1}{4} = \frac{1}{5}$ qt
6. **(2)** 8 hr $\times \frac{3}{4} = 6$ hr
7. **(2)** $5\frac{1}{4}$ hr $\times \frac{1}{3} = 1\frac{3}{4}$ hr
8. **(5)** $5\frac{1}{3}$ yd $\div \frac{2}{3}$ yd per pillow $= 8$ pillows
9. **(1)** $\frac{3}{4}$ hr $\div 9$ boxes $= \frac{1}{12}$ hr per box
10. **(1)** 100 lb $\div 12\frac{1}{2}$ lb per person $= 8$ people
11. **(3)** $1\frac{1}{2}$ pt $\times 16$ fl oz per pint $= 24$ fl oz; 24 fl oz $\div 12$ fl oz per can $= 2$ cans
12. **(4)** $\frac{22}{7} \times 10\frac{1}{2}$ yd $\times 10\frac{1}{2}$ yd $= 346\frac{1}{2}$ sq yd; $346\frac{1}{2}$ sq yd $\div 1\frac{1}{2}$ sq yd per person $= 231$ people
13. **(3)** $\$4.80 \times .75 = \$3.60$
14. **(5)** $12\frac{1}{2}$ ft $\times 16$ ft $= 200$ sq ft; 200 sq ft $\times \$3$ per sq ft $= \$600$
15. **(2)** 8 ft $\times 11\frac{1}{2}$ ft $= 92$ sq ft; 92 sq ft $\div 46$ sq ft per can $= 2$ cans

UNIT 2   PERCENTS

EXERCISE 27A (PAGE 111)

Part A

1. (a) **10%**   2. (a) **25%**
   (b) **90%**   (b) **75%**
3. (a) **15%**   4. (a) **50%**
   (b) **85%**   (b) **50%**
5. (a) **33%**   6. (a) **1%**
   (b) **67%**   (b) **99%**

Part B

1. **25%**   2. **50%**
3. **31%**   4. **100%**

Word Problem

The first night **88%** of the class was present.

EXERCISE 27B (PAGE 113)

1. **125%**   2. **160%**
3. **175%**   4. **105%**
5. **195%**   6. **115%**
7. **102%**   8. **128%**
9. **199%**   10. **200%**

Word Problem

At the end of one year Jerry had **105%** of his original deposit.

EXERCISE 28A (PAGE 117)

1. **.55**   2. **.32**   3. **.67**   4. **.73**
5. **.44**   6. **.2**   7. **.1**   8. **.88**
9. **.7**   10. **.6**   11. **.14** 12. **.9**
13. **.12** 14. **.3**   15. **.76** 16. **.99**
17. **.4**   18. **.59** 19. **.8**   20. **.41**

Word Problem

This month **.87** (eighty-seven hundredths) of Gull Wing Airways's arrivals were on time.

EXERCISE 28B (PAGE 118)

1. **.05**   2. **.08**   3. **.02**   4. **.09**
5. **.01**   6. **.07**   7. **.03**   8. **.06**

Word Problem

Harriet had **.05** (five hundredths) of her answers wrong.

EXERCISE 28C (PAGE 118)

1. **1.11**   2. **2.31**   3. **4.53**   4. **3.12**
5. **7.68**   6. **1.76**   7. **3.25**   8. **5.75**
9. **6.25**   10. **3.12** 11. **9.89** 12. **2.11**
13. **9.09** 14. **8.54** 15. **4.58** 16. **3.56**
17. **2.01** 18. **1.02** 19. **6.07** 20. **6.53**

Word Problem

Stan's account has increased **2.5** times.

EXERCISE 29A (PAGE 120)

1. **45%**   2. **67%**   3. **2%**   4. **9%**
5. **75%**   6. **7%**   7. **44%**   8. **8%**
9. **1%**   10. **33%** 11. **54%** 12. **78%**
13. **98%** 14. **12%** 15. **23%** 16. **66%**
17. **76%** 18. **89%** 19. **3%**   20. **8%**

Word Problem

José burned **11%** of the heating oil during the two weeks.

EXERCISE 29B (PAGE 121)

1. **64.5%**   2. **84.5%**   3. **7.5%**   4. **76.7%**
5. **2.4%**   6. **33.3%**   7. **28.9%**   8. **75.1%**
9. **88.9%**   10. **3.2%**   11. **29.7%** 12. **78.9%**
13. **87.65%** 14. **.43%**   15. **5.04%** 16. **11.1%**
17. **31.2%** 18. **73.55%** 19. **14.15%** 20. **7.77%**

Word Problem

An average of **44.5%** of the applicants were accepted.

EXERCISE 29C (PAGE 122)

1. **20%** 2. **50%** 3. **70%** 4. **90%**
5. **10%** 6. **40%** 7. **60%** 8. **80%**

Word Problem

Althea spends **20%** of her earnings for food.

EXERCISE 30A (PAGE 123)

1. $\frac{3}{10}$ 2. $\frac{4}{5}$ 3. $5\frac{1}{10}$ 4. $\frac{9}{10}$
5. $\frac{7}{10}$ 6. $3\frac{1}{4}$ 7. $\frac{3}{4}$ 8. $\frac{1}{2}$
9. $\frac{3}{5}$ 10. $\frac{11}{20}$ 11. $1\frac{3}{25}$ 12. $\frac{19}{20}$
13. $2\frac{2}{25}$ 14. $\frac{1}{25}$ 15. $\frac{24}{25}$ 16. $\frac{7}{20}$
17. $\frac{9}{20}$ 18. $\frac{2}{5}$ 19. $3\frac{1}{5}$ 20. $\frac{1}{20}$

Word Problem

$\frac{1}{4}$ of the students will be new to the school.

EXERCISE 30B (PAGE 124)

1. $\frac{1}{8}$ 2. $\frac{1}{200}$ 3. $\frac{3}{8}$ 4. $\frac{1}{16}$
5. $\frac{7}{8}$ 6. $\frac{1}{500}$ 7. $\frac{9}{40}$ 8. $\frac{13}{40}$
9. $1\frac{1}{8}$ 10. $3\frac{3}{8}$ 11. $\frac{1}{40}$ 12. $\frac{1}{2500}$

Word Problem

$\frac{3}{8}$ of the employees came to work.

EXERCISE 30C (PAGE 125)

1. $1\frac{1}{8}$ 2. $\frac{1}{200}$ 3. $\frac{3}{8}$ 4. $\frac{1}{16}$
5. $\frac{7}{8}$ 6. $\frac{1}{500}$ 7. $\frac{1}{1000}$ 8. $\frac{3}{400}$
9. $\frac{31}{400}$ 10. $\frac{3}{1000}$ 11. $\frac{33}{500}$ 12. $\frac{999}{1000}$

Word Problem

$\frac{17}{200}$ of the goods were defective.

EXERCISE 31A (PAGE 126)

1. **1%** 2. **17%** 3. **33%** 4. **51%** 5. **99%**

Word Problem

Alan had to shade **33%** of the grid.

EXERCISE 31B (PAGE 128)

1. **50%** 2. **6.25%** 3. **12.5%** 4. **5%**
5. **25%** 6. **75%** 7. **80%** 8. **37.5%**
9. **18.75%** 10. **40%** 11. **28%** 12. **31.25%**
13. **83$\frac{1}{3}$%** 14. **33$\frac{1}{3}$%** 15. **66$\frac{2}{3}$%** 16. **42$\frac{6}{7}$%**

Word Problem

Of all the households on Franklin Street, **93.75%** belong to the community association.

EXERCISE 32 (PAGE 130)

1. 60% $\frac{3}{5}$ .6
2. **25%** $\frac{1}{4}$ .25
3. **50%** $\frac{1}{2}$ .5
4. 75% $\frac{3}{4}$ .75
5. **10%** $\frac{1}{10}$ .1
6. **20%** $\frac{1}{5}$ .2
7. 30% $\frac{3}{10}$ .3
8. **16$\frac{2}{3}$%** $\frac{1}{6}$ .16$\frac{2}{3}$
9. 62.5% $\frac{5}{8}$ **.625**
10. **33$\frac{1}{3}$%** $\frac{1}{3}$ .33$\frac{1}{3}$
11. **12.5%** $\frac{1}{8}$ .125
12. **66$\frac{2}{3}$%** $\frac{2}{3}$ .66$\frac{2}{3}$

Word Problem

Miguel paid **33$\frac{1}{3}$%** as a down payment.

EXERCISE 33A (PAGE 132)

2. (a) 90 is the **whole.**
   (b) 4.5 is the **part.**
   (c) You must find the **percent.**

3. (a) 18 is the **part.**
   (b) 20% is the **percent.**
   (c) You must find the **whole.**

4. (a) 9 is the **part.**
   (b) 4.5 is the **whole.**
   (c) You must find the **percent.**

5. (a) 2 is the **part.**
   (b) 25% is the **percent.**
   (c) You must find the **whole.**

6. (a) 70% is the **percent.**
   (b) 15 is the **whole.**
   (c) You must find the **part.**

Word Problem

The problem asks you to find **the whole.**

EXERCISE 33B (PAGE 134)

2. (a) You must find **the percent.**
   (b) You would **divide 4.5 by 90.**
3. (a) You must find **the whole.**
   (b) You would **divide 18 by 20%.**
4. (a) You must find **the percent.**
   (b) You would **divide 9 by 4.5.**

Answers and Solutions

5. (a) You must find **the whole.**
   (b) You would **divide 2 by 25%.**
6. (a) You must find **the part.**
   (b) You would **multiply 70% times 15.**

Word Problem

To find the whole, you would **divide 15 by 75%.**

EXERCISE 34A (PAGE 136)

1. $.16 \times 150 =$ **24**
2. $.007 \times 200 =$ **1.4**
3. $.08 \times 350 =$ **28**
4. $.09 \times 400 =$ **36**
5. $.18 \times \$72 =$ **\$12.96**
6. $.015 \times 245 =$ **3.675**
7. $.11 \times \$34 =$ **\$3.74**
8. $.0014 \times \$56 =$ **\$.08**
9. $.62 \times \$11 =$ **\$6.82**
10. $.09 \times \$6 =$ **\$.54**
11. $.12 \times \$72 =$ **8.64**
12. $.072 \times \$143 =$ **\$10.30**
13. $1.15 \times 240 =$ **276**
14. $.378 \times \$1500 =$ **\$567**
15. $.0233 \times 46 =$ **1.0718**
16. $.00142 \times \$390 =$ **\$.55**
17. $2.12 \times \$9 =$ **\$19.08**
18. $.105 \times \$24.80 =$ **\$2.60**
19. $1.0875 \times \$20 =$ **\$21.75**
20. $.0875 \times \$25 =$ **\$2.19**

Word Problem

Simone saves **\$177** per month.
$.12 \times \$1475 = \$177$

EXERCISE 34B (PAGE 137)

1. $\frac{3}{5} \times 80 =$ **48**
2. $\frac{1}{2} \times 128 =$ **64**
3. $\frac{7}{10} \times 50 =$ **35**
4. $2\frac{1}{4} \times 160 =$ **360**
5. $\frac{3}{4} \times 600 =$ **450**
6. $\frac{2}{3} \times 126 =$ **84**
7. $\frac{7}{8} \times 160 =$ **140**
8. $\frac{1}{6} \times 360 =$ **60**
9. $1\frac{1}{8} \times \$98 =$ **\$110.25**
10. $\frac{3}{8} \times \$196 =$ **\$73.50**
11. $\frac{5}{6} \times \$72 =$ **\$60**
12. $1\frac{1}{10} \times \$12.80 =$ **\$14.08**
13. $\frac{51}{100} \times 1000 =$ **510**
14. $\frac{13}{100} \times \$10 =$ **\$1.30**

Word Problem

There are **15 men** in the class.
$$\frac{1}{3} \times 45 = 15$$

EXERCISE 35A (PAGE 139)

1. $9 \div 75 = .12 =$ **12%**
2. $16 \div 200 = .08 =$ **8%**
3. $102 \div 120 = .85 =$ **85%**
4. $3 \div 200 = .015 =$ **1.5%**
5. $24 \div 500 = .048 =$ **4.8%**
6. $388 \div 400 = .97 =$ **97%**
7. $4.5 \div 150 = .03 =$ **3%**
8. $.3 \div 6 = .05 =$ **5%**
9. $7 \div 400 = .0175 =$ **1.75%**
10. $40 \div 320 = .125 =$ **12.5%**

Word Problem

Manny sold **88%** of the books that week.
$$264 \div 300 = .88 = 88\%$$

EXERCISE 35B (PAGE 140)

1. $\frac{30}{40} = \frac{3}{4} =$ **75%**
2. $\frac{8}{16} = \frac{1}{2} =$ **50%**
3. $\frac{7}{70} = \frac{1}{10} =$ **10%**
4. $\frac{4}{20} = \frac{1}{5} =$ **20%**
5. $\frac{25}{25} = \frac{1}{1} =$ **100%**
6. $\frac{30}{90} = \frac{1}{3} =$ **33$\frac{1}{3}$%**
7. $\frac{25}{200} = \frac{1}{8} =$ **12$\frac{1}{2}$%**
8. $\frac{50}{75} = \frac{2}{3} =$ **66$\frac{2}{3}$%**
9. $\frac{20}{32} = \frac{5}{8} =$ **62$\frac{1}{2}$%**
10. $\frac{9}{18} = \frac{1}{2} =$ **50%**
11. $\frac{13}{78} = \frac{1}{6} =$ **16$\frac{2}{3}$%**
12. $\frac{40}{48} = \frac{5}{6} =$ **83$\frac{1}{3}$%**
13. $\frac{98}{200} = \frac{49}{100} =$ **49%**
14. $\frac{39}{300} = \frac{13}{100} =$ **13%**
15. $\frac{70}{1000} = \frac{7}{100} =$ **7%**
16. $\frac{250}{4000} = \frac{1}{16} =$ **6$\frac{1}{4}$%**

Word Problem

The down payment was **20%** of the purchase price.
$$\frac{300}{1500} = \frac{1}{5} = 20\%$$

EXERCISE 35C (PAGE 141)

1. $\frac{24}{8} = 3 =$ **300%**
2. $\frac{25}{1.25} = 20 =$ **2000%**
3. $\frac{6}{3} = 2 =$ **200%**
4. $\frac{12}{4} = 3 =$ **300%**
5. $\frac{14}{4} = 3.5 =$ **350%**
6. $\frac{2}{1} = 2 =$ **200%**
7. $\frac{156}{4} = 39 =$ **3900%**
8. $\frac{6}{4} = 1.5 =$ **150%**
9. $\frac{75}{50} = 1.5 =$ **150%**
10. $\frac{36}{27} = 1\frac{1}{3} =$ **133$\frac{1}{3}$%**
11. $\frac{88}{55} = 1\frac{3}{5} =$ **160%**
12. $\frac{75}{12.5} = 6 =$ **600%**

Word Problem

This week's receipts were **125%** of last week's receipts.
$$600 \div 480 = 1.25 = 125\%$$
or
$$\frac{600}{480} = \frac{5}{4} = 1\frac{1}{4} = 125\%$$

EXERCISE 36A (PAGE 143)

1. 840 ÷ .0025 = **336,000**
2. 5 ÷ .02 = **250**
3. 10 ÷ .01 = **1000**
4. 34.5 ÷ .23 = **150**
5. 8 ÷ .04 = **200**
6. 21 ÷ .07 = **300**
7. 15 ÷ .03 = **500**
8. $11.75 ÷ .47 = **$25**
9. $78.75 ÷ .0875 = **$900**
10. $59.50 ÷ 1.19 = **$50**
11. $7.20 ÷ 1.28 = **$5.63**
12. .019 ÷ .38 = **.05**
13. .42 ÷ .21 = **2**
14. 1.6 ÷ .48 = **3.33$\frac{1}{3}$**
15. 99.44 ÷ .9944 = **100**
16. $210 ÷ 1.05 = **$200**

Word Problem

Kenia's salary is **$24,000.**

$$3840 \div .16 = \$24,000$$

EXERCISE 36B (PAGE 144)

1. $8 \div \frac{1}{4} = 8 \times \frac{4}{1} =$ **32**
2. $2 \div \frac{1}{8} = 2 \times \frac{8}{1} =$ **16**
3. $36 \div \frac{3}{10} = 36 \times \frac{10}{3} =$ **120**
4. $15 \div \frac{1}{3} = 15 \times \frac{3}{1} =$ **45**
5. $48 \div \frac{3}{8} = 48 \times \frac{8}{3} =$ **128**
6. $49 \div \frac{7}{8} = 49 \times \frac{8}{7} =$ **56**
7. $8 \div \frac{1}{8} = 8 \times \frac{8}{1} =$ **64**
8. $64 \div \frac{2}{5} = 64 \times \frac{5}{2} =$ **160**
9. $48 \div \frac{2}{3} = 48 \times \frac{3}{2} =$ **72**
10. $\$50 \div 1\frac{1}{4} = \$50 \times \frac{4}{5} =$ **$40**
11. $500 \div 2 = 500 \times \frac{1}{2} =$ **250**
12. $12 \div 1\frac{1}{5} = 12 \times \frac{5}{6} =$ **10**
13. $240 \div 1\frac{1}{2} = 240 \times \frac{2}{3} =$ **160**
14. $840 \div 1\frac{1}{3} = 840 \times \frac{3}{4} =$ **630**
15. $100 \div 2\frac{1}{2} = 100 \times \frac{2}{5} =$ **40**
16. $78 \div \frac{39}{100} = 78 \times \frac{100}{39} =$ **200**
17. $57 \div \frac{19}{100} = 57 \times \frac{100}{19} =$ **300**
18. $51 \div \frac{51}{100} = 51 \times \frac{100}{51} =$ **100**

Word Problem

Altogether there are **90** merchants on Pine Avenue.

$$63 \div \frac{7}{10} = 63 \times \frac{10}{7} = 90$$

PERCENT SKILLS REVIEW (page 146)

Part A

1. **23%**
2. **77%**
3. **150%**
4. **40%**

Part B

1. **240%**   2. **62.5%**   3. **66$\frac{2}{3}$%**   4. **37$\frac{1}{2}$%**
   or **62$\frac{1}{2}$%**
5. **.6%**   6. **33.3%**   7. **5%**   8. **83$\frac{1}{3}$%**
9. **7006.5%**   10. **29.9%**

Part C

1. **.032**   2. **.42**   3. **.0006**   4. **.05**
5. **1.09**   6. **.125**   7. **.16$\frac{2}{3}$**   8. **.2**
9. **3.33$\frac{1}{3}$**   10. **.5**

Part D

1. $\frac{7}{8}$ 2. $\frac{5}{6}$ 3. $1\frac{1}{8}$ 4. $\frac{3}{5}$
5. $\frac{3}{4}$ 6. $\frac{1}{3}$ 7. $\frac{1}{2}$ 8. $4\frac{3}{10}$
9. $\frac{1}{100}$ 10. $1\frac{1}{5}$

Part E

1. .75 × 80 = **60**   2. $\frac{7}{28} = \frac{1}{4} =$ **25%**
3. .005 × $92 = **$.46**   4. $\frac{\$80}{\$8} = 10 =$ **1000%**
5. $\frac{2}{3} \times 117 =$ **78**   6. 2.2 × 500 = **1100**
7. 120 ÷ 1.5 = **80**   8. .12 × 30 = **3.6**
9. 206 ÷ 2 = **103**   10. .0135 × 150 = **2.025**
11. 16 ÷ 25.6 = .625 = **62.5%**
12. 27 ÷ .09 = **300**
13. 12 ÷ .015 = **800**
14. .00125 × 12,800 = **16**
15. 90 ÷ 45 = 2 = **200%**   16. $\frac{\$45}{\$90} = \frac{1}{2} =$ **50%**
17. 1 × 9.8 = **9.8**   18. $\frac{20}{32} = \frac{5}{8} =$ **62$\frac{1}{2}$%**
19. $35 ÷ .7 = **$50**   20. $\frac{8}{48} = \frac{1}{6} =$ **16$\frac{2}{3}$%**

21. $6 \div \frac{3}{8} = $ **16** 22. $\$16 \div \$5 = 3.2 = $ **320%**
23. $45 \div \frac{1}{3} = $ **135** 24. $10 \times .075 = $ **.75**
25. $54 \div .6 = $ **90**

EXERCISE 37A (PAGE 149)

The following solutions show one way to estimate each answer. You may have estimated differently.

1. 206 can be changed to 200: $\frac{1}{2} \times 200 = $ **100.**
2. 38 can be changed to 39: $\frac{1}{3} \times 39 = $ **13.**
3. 45 can be changed to 44: $\frac{1}{4} \times 44 = $ **11.**
4. 98 can be changed to 100: $\frac{7}{8} \times 100 = $ **$87\frac{1}{2}$.**
5. 322 can be changed to 300: $.4 \times 300 = $ **120.**
6. 325 can be changed to 320: $\frac{1}{8} \times 320 = $ **40.**
7. 231 can be changed to 200: $.7 \times 200 = $ **140.**
8. 65 can be changed to 64: $\frac{3}{8} \times 64 = $ **24.**
9. 38 can be changed to 36: $\frac{1}{6} \times 36 = $ **6.**
10. 103 can be changed to 100: $\frac{1}{20} \times 100 = $ **5.**

Word Problem

Evita spent about **$6000** on rent and utilities last year.

$17,897 can be changed to $18,000.

$$\frac{1}{3} \times \$18,000 = \$6000$$

EXERCISE 37B (PAGE 150)

The following solutions show one way to estimate each answer. You may have estimated differently.

1. 74% can be changed to 75%: $\frac{3}{4} \times 400 = $ **300.**
2. 51% can be changed to 50%: $\frac{1}{2} \times 150 = $ **75.**
3. 9% can be changed to 10%: $\frac{1}{10} \times 300 = $ **30.**
4. 47% can be changed to 50%: $\frac{1}{2} \times 86 = $ **43.**
5. 33% can be changed to $33\frac{1}{3}$%: $\frac{1}{3} \times 999 = $ **333.**
6. 26% can be changed to 25%: $\frac{1}{4} \times 80 = $ **20.**
7. 34% can be changed to $33\frac{1}{3}$%: $\frac{1}{3} \times 18 = $ **6.**
8. 13% can be changed to $12\frac{1}{2}$%: $\frac{1}{8} \times 16 = $ **2.**
9. 17% can be changed to $16\frac{2}{3}$%: $\frac{1}{6} \times 72 = $ **12.**
10. 39% can be changed to 40%: $\frac{2}{5} \times 50 = $ **20.**

Word Problem

About **400** applicants failed the physical.

47% can be changed to 50%.

$$\frac{1}{2} \times 800 = 400$$

EXERCISE 37C (PAGE 152)

The following solutions show one way to estimate each answer. You may have estimated differently.

1. 32% can be changed to $33\frac{1}{3}$%: $15 \div \frac{1}{3} = $ **45.**
2. 27% can be changed to 25%: $20 \div \frac{1}{4} = $ **80.**
3. 29 can be changed to 30: $30 \div \frac{1}{3} = $ **90.**
4. 11% can be changed to 10%: $6.2 \div .1 = $ **62.**
5. 4.1 can be changed to 4: $4 \div \frac{1}{8} = $ **32.**
6. 24% can be changed to 25%: $8 \div \frac{1}{4} = $ **32.**
7. 9% can be changed to 10%: $116 \div \frac{1}{10} = $ **1160.**
8. 362 can be changed to 360: $360 \div .9 = $ **400.**
9. 19% can be changed to 20%: $20 \div \frac{1}{5} = $ **100.**
10. 1.7 can be changed to 2: $2 \div \frac{1}{6} = $ **12.**

Word Problem

There are about **160** employees at the mall.

24.7% can be changed to 25%.

$$40 \div \frac{1}{4} = 160$$

EXERCISE 37D (PAGE 153)

The following solutions show one way to estimate each answer. You may have estimated differently.

1. 29 can be changed to 30: $\frac{30}{40} = \frac{3}{4} = $ **75%.**
2. 11 can be changed to 12: $\frac{12}{16} = \frac{3}{4} = $ **75%.**
3. 69 can be changed to 70: $\frac{7}{70} = \frac{1}{10} = $ **10%.**
4. 21 can be changed to 20: $\frac{4}{20} = \frac{1}{5} = $ **20%.**
5. 249 can be changed to 250: $\frac{250}{250} = 1 = $ **100%.**
6. 89 can be changed to 90: $\frac{30}{90} = \frac{1}{3} = $ **$33\frac{1}{3}$%.**
7. 21 can be changed to 20: $\frac{20}{200} = \frac{1}{10} = $ **10%.**
8. 23 can be changed to 25: $\frac{25}{75} = \frac{1}{3} = $ **$33\frac{1}{3}$%.**
9. 21 can be changed to 20: $\frac{20}{32} = \frac{5}{8} = $ **$62\frac{1}{2}$%.**
10. 19 can be changed to 18: $\frac{9}{18} = \frac{1}{2} = $ **50%.**

Word Problem

About **20%** of John's customers used food stamps.

61 can be changed to 60.

$$\frac{12}{60} = \frac{1}{5} = 20\%$$

EXERCISE 38 (PAGE 156)

1. The tax was **$3.60.**
   (a) What is 6% of $510?
   (b) part = % × whole
   (c) part = .06 × $500 = $30
   (d) part = .06 × $510 = $3.60

2. The sales tax was **9%.**
   (a) $45 is what percent of $500?
   (b) % = part ÷ whole
   (c) % = $\frac{\$50}{\$500} = \frac{1}{10} = 10\%$
   (d) % = $45 ÷ $500 = .09 = 9%

3. The daily rate rose by **$22\frac{2}{9}$%.**
   (a) $1 is what percent of $4.50?
   (b) % = part ÷ whole
   (c) % = $\frac{\$1}{\$5} = 20\%$
   (d) % = $1 ÷ $4.50 = $.22\frac{2}{9} = 22\frac{2}{9}\%$

4. Before Christmas the jackets had cost **$125.**
   (a) $85 is 68% of what amount?
   (b) whole = part ÷ %
   (c) whole = $84 ÷ .7 = 120
   (d) whole = $85 ÷ .68 = $125

5. The earlier speed limit had been **45 miles per hour.**
   (a) 30 is $66\frac{2}{3}$% of what number?
   (b) whole = part ÷ %
   (c) whole = 30 mph ÷ .6 = 50 mph
   (d) whole = 30 mph ÷ $\frac{2}{3}$ = 45 mph

6. The bank paid interest at a rate of **5.75%** on her deposit.
   % = part ÷ whole
   % = $287.50 ÷ $5000
   % = .0575 = 5.75%

7. The store charged **$4.50** for the calendars.
   part = % × whole
   part = 3.75 × $1.20
   part = $4.50

8. She paid **$12,360** for the car when it was new.
   whole = part ÷ %
   whole = $2060 ÷ $\frac{1}{6}$
   whole = $12,360

9. The price of the rocking chair before the sales tax was added was **$300.**
   whole = part ÷ %
   whole = $24 ÷ .08
   whole = $300

10. The store marks up the price it pays for shirts by **$66\frac{2}{3}$%.**
    % = part ÷ whole
    % = $\frac{\$8}{\$12}$
    % = $\frac{2}{3} = 66\frac{2}{3}\%$

11. Kwon Li paid **$126.88** sales tax on the living room set.
    part = % × whole
    part = .0875 × $1450
    part = $126.88

12. The value of the comic book had appreciated **$22.50** by the time Jerry sold it.
    part = % × whole
    part = .375 × $60
    part = $22.50

13. Franco had to pay **$198** interest on the money he borrowed.
    part = % × whole
    part = .11 × $1800
    part = $198

14. The sales tax was **6%.**
    % = part ÷ whole
    % = $856.80 ÷ $14,280
    % = .06 = 6%

15. The lottery winner had won **$1,200,000.**
    whole = part ÷ %
    whole = $588,000 ÷ .49
    whole = $1,200,000

EXERCISE 39A (PAGE 159)

1. Including the tax, Tanya paid **$96.75** for the sweaters.
   100% + 7.5% = 107.5%
   part = % × whole
   part = 1.075 × $90
   part = $96.75

2. The temperature was **72°** at 3:00 P.M.
   100% + 140% = 240%
   part = % × whole
   part = 2.4 × 30°
   part = 72°

Answers and Solutions   239

3. The new car would get **42 miles per gallon.**
   100% + 20% = 120%
   part = % × whole
   part = 1.2 × 35 miles per gallon
   part = 42 miles per gallon

4. Adrienne charged **$115.63** for each dress.
   100% + 85% = 185%
   part = % × whole
   part = 1.85 × $62.50
   part = $115.625, which rounds to $115.63

5. Angelo charged **$12.53** for a case of the motor oil.
   100% + 40% = 140%
   part = % × whole
   part = 1.4 × $8.95
   part = $12.53

6. Larry sold the motorcycle for **$797.50.**
   100% + 45% = 145%
   part = % × whole
   part = 1.45 × $550
   part = $797.50

7. Marcia's salary will be **$19,740** after the increase.
   100% + 5.28% = 105.28%
   part = % × whole
   part = 1.0528 × $18,750
   part = $19,740

8. Toby's necklace is now worth **$420.**
   100% + 12% = 112%
   part = % × whole
   part = 1.12 × $375
   part = $420

9. The value of the money Marie added was **$552.08** after a year.
   $100\% + 8\frac{1}{4}\% = 108\frac{1}{4}\%$
   part = % × whole
   part = 1.0825 × $510
   part = $552.075, which rounds to $552.08

10. James's total income for the year was **$42,000.**
    $100\% + 16\frac{2}{3}\% = 116\frac{2}{3}\%$
    part = % × whole
    part = $1\frac{1}{6}$ × $36,000
    part = $42,000

EXERCISE 39B (PAGE 161)

1. After a year, the bulldozer's value was **$127,500.**
   100% − 15% = 85%
   part = % × whole
   part = .85 × $150,000
   part = $127,500

2. After its value fell, the co-op was worth **$74,000.**
   $100\% - 33\frac{1}{3}\% = 66\frac{2}{3}\%$
   part = % × whole
   part = $\frac{2}{3}$ × $111,000
   part = $74,000

3. Sylvia will pay **$427.50** for the word processor.
   100% − 25% = 75%
   part = % × whole
   part = .75 × $570
   part = $427.50

4. The sale price for one of last year's cars was **$10,040.**
   100% − 20% = 80%
   part = % × whole
   part = .8 × $12,550
   part = $10,040

5. The sofa's sale price was **$465.60.**
   100% − 40% = 60%
   part = % × whole
   part = .6 × $776
   part = $465.60

6. The discounted fare was **$681.25.**
   $100\% - 37\frac{1}{2}\% = 62\frac{1}{2}\%$
   part = % × whole
   part = .625 × $1090
   part = $681.25

7. The sale price of the appointment books was **$7.50.**
   100% − 50% = 50%
   part = % × whole
   part = $\frac{1}{2}$ × $15
   part = $7.50

8. Juan's salary was **$21,560** after the cut.
   100% − 12% = 88%
   part = % × whole
   part = .88 × $24,500
   part = $21,560

9. Blanca received **$1015** after the collapse.
   100% − 65% = 35%
   part = % × whole
   part = .35 × $2900
   part = $1015

10. A month after she bought it, Mary's stock was worth **$30.77** per share.
    100% − 11% = 89%
    part = % × whole
    part = .89 × $34.57
    part = $30.7673, which rounds to $30.77

EXERCISE 39C (PAGE 163)

1. The bookstore marked the almanacs up by **376%**.
   $5.95 − $1.25 = $4.70
   % = part ÷ whole
   % = $4.70 ÷ $1.25
   % = 3.76 = 376%

2. Peggy had marked the table up by **50%**.
   $360 − $240 = $120
   % = part ÷ whole
   % = $120 ÷ $240
   % = .5 = 50%

3. Peggy reduced the cost of the painting by **20%**.
   $100 − $80 = $20
   % = part ÷ whole
   % = $20 ÷ $100
   % = .2 = 20%

4. Audrey realized a **10%** profit when she sold the land.
   $7920 − $7200 = $720
   % = part ÷ whole
   % = $720 ÷ $7200
   % = .1 = 10%

5. Allen marked the tomatoes up by **150%**.
   $5.50 − $2.20 = $3.30
   % = part ÷ whole
   % = $3.30 ÷ $2.20
   % = 1.5 = 150%

6. Allen marked down the tomatoes below his cost by **50%**.
   $2.20 − $1.10 = $1.10
   % = part ÷ whole
   % = $1.10 ÷ $2.20
   % = .5 = 50%

7. Your coin's value has increased by **200%** since you bought it.
   $750 − $250 = $500
   % = part ÷ whole
   % = $500 ÷ $250
   % = 2 = 200%

8. Peter marked up his cost for the scallops by $16\frac{2}{3}$%.
   $3.92 − $3.36 = $.56
   % = part ÷ whole
   % = $.56 ÷ $3.36
   % = .1667 = $16\frac{2}{3}$%

9. The store had discounted the exercise shoes by **35%**.
   $60 − $39 = $21
   % = part ÷ whole
   % = $21 ÷ $60
   % = .35 = 35%

10. Joaquin's wage has gone up by **5%**.
    $12.18 − $11.60 = $.58
    % = part ÷ whole
    % = $.58 ÷ $11.60
    % = .05 = 5%

PERCENTS REVIEW (page 164)

Part A

1. **75%**
2. **25%**
3. **515%**
4. **90%**
5. **10%**

Part B

1. **3%** 2. **8.25%** 3. **25%** 4. **20%** 5. **37.5%** or $37\frac{1}{2}$%

Part C

1. 50%	$\frac{1}{2}$	.5
2. 75%	$\frac{3}{4}$	.75
3. $33\frac{1}{3}$%	$\frac{1}{3}$	$.33\frac{1}{3}$
4. 87.5%	$\frac{7}{8}$	.875
5. $83\frac{1}{3}$%	$\frac{5}{6}$	$.83\frac{1}{3}$
6. 20%	$\frac{1}{5}$	.2
7. $16\frac{2}{3}$%	$\frac{1}{6}$	$.16\frac{2}{3}$
8. 62.5%	$\frac{5}{8}$	.625
9. 10%	$\frac{1}{10}$	.1
10. 12.5%	$\frac{1}{8}$	.125

Part D

1. .60 × 80 = **48**
2. $\frac{9}{36} = \frac{1}{4}$ = **25%**
3. .5 × 128 = **64**
4. $\frac{\$17.50}{\$35} = \frac{1}{2}$ = **50%**
5. .16 × 150 = **24**
6. $\frac{2}{3}$ × 96 = **64**
7. 9 ÷ $\frac{1}{6}$ = **54**
8. .005 × 400 = **2**
9. 25 ÷ 2.5 = **10**
10. 100 ÷ 1,000,000 = .0001 = **.01%**

Part E

1. 5983 can be changed to 6000.
   $\frac{2000}{6000} = \frac{1}{3} = 33\frac{1}{3}\%$
   This solution shows one way to estimate the answer. You may have estimated differently.

2. Melanie's commission on the sale was **$38.88**.
   part = % × whole
   part = .09 × $432
   part = $38.88

3. The suits were marked down by **20%**.
   % = part ÷ whole
   % = $25 ÷ $125
   % = .2 = 20%

4. Falcon paid **$16,000** for the car.
   whole = part ÷ %
   whole = $12,000 ÷ .75
   whole = $16,000

5. There were **96** traffic deaths this year.
   100% − 36% = 64%
   part = % × whole
   part = .64 × 150
   part = 96

6. Jeremy paid **$16,312.50** altogether for the car.
   100% + 8.75% = 108.75%
   part = % × whole
   part = 1.0875 × $15,000
   part = $16,312.50

7. Li marks up his cost for the calendars by **$233\frac{1}{3}\%$**.
   $5.00 − $1.50 = $3.50
   % = part ÷ whole
   % = $3.50 ÷ $1.50
   % = $2.33\frac{1}{3} = 233\frac{1}{3}\%$

8. The store had discounted the paint by **20%**.
   $15 − $12 = $3
   % = part ÷ whole
   % = $3 ÷ $15
   % = .2 = 20%

9. George had $83\frac{1}{3}\%$ of the problems right.
   % = part ÷ whole
   % = $\frac{15}{18} = \frac{5}{6}$
   % = $.83\frac{1}{3} = 83\frac{1}{3}\%$

10. Felix's new salary will be **$14,568.13**.
    100% + 5.95% = 105.95%
    part = % × whole
    part = 1.0595 × $13,750
    part = $14,568.125, which rounds to $14,568.13

GED PRACTICE 2 (page 167)

1. (2) part = % × whole
   part = .075 × $180
   part = $13.50

2. (1) % = part ÷ whole
   % = $227.50 ÷ $3500
   % = .065 = 6.5%

3. (4) whole = part ÷ %
   whole = $128 ÷ .08
   whole = $1600.00

4. (5) % = part ÷ whole
   % = 120 sq ft ÷ 192 sq ft
   % = .625 = 62.5%

5. (5) part = % × whole
   part = .2 × 45 students
   part = 9 students

6. (3) whole = part ÷ %
   whole = $1000.00 ÷ $\frac{1}{6}$
   whole = $6000.00

7. (2) % = part ÷ whole
   % = $450 ÷ $9000
   % = .05 = 5%

8. (4) % = part ÷ whole
   % = 28 in. ÷ 200 in.
   % = 1.4 = 140%

9. (2) 149,763 fans can be changed to 150,000 fans.
   100% + 50% = 150%
   part = % × whole
   part = 1.5 × 150,000 fans
   part = 225,000 fans

10. (5) 100% + 150% = 250%
    part = % × whole
    part = 2.5 × $.20
    part = $.50

11. (2) $8.50 − $8.00 = $.50
    % = part ÷ whole
    % = $.50 ÷ $8.00
    % = .0625 = 6.25%

12. (4) $.79 can be changed to $.80.
    % = part ÷ whole
    % = $\frac{\$.10}{\$.80}$
    % = $\frac{1}{8}$ = 12.5%

13. (4) 100% − 20% = 80%
    part = % × whole
    part = .8 × $280
    part = $224

14. **(5)** 250 lb − 230 lb = 20 lb
% = part ÷ whole
% = 20 lb ÷ 250 lb
% = .08 = 8%

15. **(4)** 100% + 4% = 104%
part = % × whole
part = .04 × $4800
part = $4992

UNIT 3   TABLES AND GRAPHS

EXERCISE 40A (PAGE 175)

1. The table shows information about **the causes of accidental deaths in the U.S.A.**
2. The information in the table came from the **National Safety Council.**
3. There were **1,900** fatal gun accidents in 1988.
4. **Motor vehicle accidents** resulted in the most deaths during each year shown in the table.
5. More than 12,000 fatalities were caused by accidental falls in **1987.**
6. Poisons caused 4,300 deaths in **1987** and **1989.**
7. Accidents involving **guns** resulted in the fewest deaths during each year shown in the table.
8. Fewer than 5,000 people were killed by burns in **1990.**
9. In 1990, **3,600** people died by choking.
10. **Drowning** resulted in 5,600 deaths in 1989.

EXERCISE 40B (PAGE 179)

1. The graph reports information about **numbers of GED graduates.**
2. The information in the graph came from **a report titled *The County GED Program: Success Rate* prepared by the Middle County School Board.**
3. There were the fewest GED graduates in **1986.**
4. There were the most GED graduates in **1991.**
5. The number *18* on the scale stands for **180 graduates.**
6. There were about **330** graduates in 1987.
7. There were about **520** graduates in 1990.
8. There were about 450 graduates in **1988.**
9. The increase in the number of graduates was the greatest **between 1986 and 1987.**
10. The increase in the number of graduates was the smallest **between 1989 and 1990.**

EXERCISE 40C (PAGE 181)

1. The graph reports **the production of hogs in 5 counties in 1991.**
2. The information in the graph came from the **State Agricultural Commission.**
3. The labels on the rows in the graph stand for **counties.**
4. **Davis County** had the lowest hog production.
5. **Lincoln County** had the highest hog production.
6. **Grant County** and **Lee County** produced about the same number of hogs.
7. Each symbol in the graph stands for **10,000** hogs.
8. About **30,000** hogs were produced in Davis County.
9. About **35,000** hogs were produced in Cook County.
10. About 55,000 hogs were produced in **Lincoln County.**

EXERCISE 41A (PAGE 183)

1. The graph reports **the annual home-run records in the American League from 1976 through 1985.**
2. Each number on the vertical scale stands for **a number of home runs.**
3. Each number on the horizontal scale stands for **a year.**
4. The record number of home runs was greatest in **1978.**
5. From 1976 to 1978, the graph shows an **upward** trend.
6. The graph shows **a downward trend** from 1978 to 1981.
7. The vertical scale uses increments of **3.**
8. The record was 22 home runs in **1981.**
9. In 1980, **41** home runs set the record.
10. Thirty-nine home runs was the record in **1977, 1982,** and **1983.**

EXERCISE 41B (PAGE 185)

1. Before Laura began to exercise, she could do **3 sit-ups.**
2. The graph shows **an upward trend.**
3. Laura did **23 sit-ups** on the thirtieth day of her exercise program.
4. Laura probably did **8 sit-ups** on the fifth day of her exercise program.
5. Laura was probably able to do 29 sit-ups on the **thirty-fifth day** of her exercise program.

EXERCISE 42A (PAGE 187)

1. THE GRAPH SHOWS THE **SOURCES OF ENERGY IN THE U.S.A.**
2. THE INFORMATION IN THE GRAPH COME FROM THE **U.S. DEPARTMENT OF ENERGY.**
3. THE WHOLE CIRCLE STANDS FOR **ALL THE SOURCES OF ENERGY USED IN THE U.S.A.**
4. **PETROLEUM** IS THE LARGEST SOURCE OF ENERGY USED IN THE UNITED STATES.
5. **22.2%** OF U.S. ENERGY COMES FROM BURNING COAL.
6. **NUCLEAR ENERGY** SUPPLIES 6.9% OF THE ENERGY USED IN THE UNITED STATES.
7. THE GRAPH **DOES NOT SHOW** WHAT PERCENT OF THE ENERGY USED IN THE UNITED STATES IS SOLAR. IT SHOWS ONLY THAT HYDRO, SOLAR, WIND, AND OTHER SOURCES OF ENERGY COMBINED ACCOUNT FOR 8.1% OF ALL ENERGY SOURCES.

EXERCISE 42B (PAGE 189)

1. The graph shows **the population of the United States by race.**
2. The information for the graph was gathered **in 1990.**
3. The **Bureau of the Census** of the **U.S. Department of Commerce** supplied the information for the graph.
4. The whole bar stands for **the entire U.S. population.**
5. The whole bar represents **100%.**
6. The group of persons who are **American Indian, Eskimo, or Aleut** makes up the smallest segment of the U.S population.
7. The group of persons who are **White** makes up the largest segment of the U.S. population.
8. **12.1%** of the U.S. population is made up of persons who are black.
9. The group of persons who are **Asian or Pacific Islander** makes up 2.9% of the U.S. population.
10. **3.9%** of the U.S. population is categorized as *Other*.

EXERCISE 43A (PAGE 193)

1. The graph gives **population data** about U.S. cities.
2. The data are reported for **1980** and **1990.**
3. **Eight** cities are included in the graph.
4. The **U.S. Bureau of the Census** supplied the data for the graph.
5. The number *1000* on the scale stands for **1,000,000.**
6. About **7,300,000** people lived in New York, the U.S. city that was the largest in 1990.
7. The populations of **Chicago, Detroit,** and **Philadelphia** decreased between 1980 and 1990.
8. The populations of **Dallas** and **San Diego** were less than one million in 1980.
9. The population of **Los Angeles** increased the most between 1980 and 1990.
10. The population of **Houston** changed the least between 1980 and 1990.

EXERCISE 43B (PAGE 196)

1. The graph gives data about **births and deaths in the United States.**
2. The data reported cover **1955 through 1990.**
3. The **National Center for Health Statistics** of the **U.S. Department of Health and Human Services** supplied the data for the graph.
4. The number *1500* on the scale stands for **1,500,000.**
5. The increment between the numbers on the scale is **300.**
6. The line that stands for deaths shows **an upward trend.**
7. The direction of the trend of the line that stands for births changes in **1960** and in **1975.**
8. After 1975 **the number of births** increased more sharply than the number of deaths.
9. There were about **4,200,000** births in 1990.
10. There were about **2,100,000** deaths in 1990.
11. The numbers of deaths and births were closest to each other in **1975.**
12. The number of births changed least between **1965** and **1970.**

EXERCISE 44A (PAGE 199)

PART A

1. THE OPEC NATIONS PRODUCED **ABOUT $23\frac{1}{2}$ MILLION (23,500,000) BARRELS** OF CRUDE OIL PER DAY IN 1990.
2. ABOUT $11\frac{1}{2}$ **MILLION (11,500,000) BARRELS** OF CRUDE OIL WERE PRODUCED IN NORTH AMERICA PER DAY IN 1990.

```
  1,500,000 barrels produced in Canada
  2,500,000 barrels produced in Mexico
+ 7,500,000 barrels produced in the U.S.
 11,500,000 barrels produced in North America
```

244 Answers and Solutions

3. In 1990, the U.S.S.R. produced **about 4 million (4,000,000) more barrels** of crude oil per day than the United States.

   11,500,000 barrels produced in the U.S.S.R.
   − 7,500,000 barrels produced in the U.S.
   ‾‾‾‾‾‾‾‾‾‾‾‾‾‾‾‾‾‾‾‾‾‾‾‾‾‾‾‾‾‾‾‾‾‾‾‾‾‾‾
   4,000,000 more barrels produced by the U.S.S.R.

4. Throughout the world, **about 61 million (61,000,000) barrels** of crude oil were produced each day in 1990.

   1,500,000 barrels produced in Canada
   3,000,000 barrels produced in China
   2,500,000 barrels produced in Mexico
   23,500,000 barrels produced in the OPEC nations
   2,000,000 barrels produced in the U.K.
   7,500,000 barrels produced in the U.S.
   11,500,000 barrels produced in the U.S.S.R.
   + 9,500,000 barrels produced in other countries
   ‾‾‾‾‾‾‾‾‾‾‾‾‾‾‾‾‾‾‾‾‾‾‾‾‾‾‾‾‾‾‾‾‾‾‾‾‾‾‾
   61,000,000 barrels produced throughout the world

5. **China produced about $\frac{1}{2}$ million (500,000) more barrels** of crude oil per day in 1990 than Mexico.

   3,000,000 barrels produced in China
   − 2,500,000 barrels produced in Mexico
   ‾‾‾‾‾‾‾‾‾‾‾‾‾‾‾‾‾‾‾‾‾‾‾‾‾‾‾‾‾‾‾‾‾‾‾‾‾‾‾
   500,000 more barrels produced in China

Part B

1. **Union women** earned more per week than non-union women during the years shown on the graph. All the points on the line that stands for union women are higher than the points on the line that stands for non-union women.

2. From 1983 to 1986, the difference between union and non-union women's weekly earnings **increased.** Each year, there is more distance between the points in each pair of dots on the line. The graph shows that in 1983 the difference was $70; in 1984, $80; in 1985, $90; and in 1986, $100.

3. The difference between union and non-union women's weekly earnings was **$80** in 1984.
   $330 − $250 = $80

4. From 1986 to 1988, the difference between union and non-union women's weekly earnings **stayed the same.** The graph shows that in each year the difference was $100.

5. The difference between union and non-union women's weekly earnings was **$100** in 1988.
   $400 − $300 = $100

EXERCISE 44B (PAGE 202)

1. Space heating costs **$972** each year.
   $1800 × .54 = $972

2. Harry's annual cost for energy is **$1111.00**.
   $599.94 ÷ .54 = $1111.00

3. Tanya pays **$10.35** for her appliances per month.
   $45 × .23 = $10.35

4. Latif's electricity charges might have been **$24** lower.
   $480 × .05 = $24

5. Americans spent **$97,750,000,000** altogether on household energy in 1987.
   $17,595,000,000 ÷ .18 = $97,750,000,000

EXERCISE 45 (PAGE 204)

1. A 150-pound person would use **145 calories** bicycling at 5.5 mph for 20 minutes and then playing volleyball for 15 minutes.
   Bicycling: 4.7 calories × 20 minutes
       = 94 calories
   Volleyball: 3.4 calories × 15 minutes
       = 51 calories
   Both: 94 calories + 51 calories
       = 145 calories

2. Myrna used **209 calories** by running at 7.5 mph for 10 minutes, swimming for 10 minutes, and doing calisthenics for 10 minutes.
   Running: 11.3 calories × 10 minutes
       = 113 calories
   Swimming: 5.7 calories × 10 minutes
       = 57 calories
   Calisthenics: 3.9 calories × 10 minutes
       = 39 calories
   All three: 113 calories + 57 calories
       + 39 calories = 209 calories

3. The heavier man used **33 calories more** than the lighter man.
   220-pound man:
       5.0 calories × 30 minutes = 150 calories
   170-pound man:
       3.9 calories × 30 minutes = 117 calories
   Difference:
       150 calories − 117 calories = 33 calories

4. Frank would have to walk **25 minutes** at 4 mph to use up 195 calories.
   A 200-pound man uses 7.8 calories per minute walking at 4 mph
   195 calories ÷ 7.8 calories per minute
       = 25 minutes

Answers and Solutions

5. The population of the American colonies increased by **2050%** between 1670 and 1770.
   Population in 1670: 100,000
   Population in 1770: 2,150,000
   Difference: 2,150,000 − 100,000 = 2,050,000
   Percent of increase:
   2,050,000 ÷ 100,000 = 20.5 = 2050%

6. The population of the American colonies increased by less than 100% **between 1700 and 1720.**
   1700–1720: 80% population increase
   Population in 1700: 250,000
   Population in 1720: 450,000
   Difference: 450,000 − 250,000 = 200,000
   Percent of increase:
   200,000 ÷ 250,000 = .8 = 80%
   1720–1740: 100% population increase
   Population in 1720: 450,000
   Population in 1740: 900,000
   Difference: 900,000 − 450,000 = 450,000
   Percent of increase:
   450,000 ÷ 450,000 = 1 = 100%

7. The population of immigrants from Mexico increased by **55.6%** between 1980 and 1990.
   Population in 1980: 9,000,000
   Population in 1990: 14,000,000
   Difference:
   14,000,000 − 9,000,000 = 5,000,000
   Percent of increase:
   5,000,000 ÷ 9,000,000 = .556 = 55.6%

8. There were **8 million** more immigrants of Hispanic origin in the United States in 1990 than in 1980.
   Total immigrants of Hispanic origin in 1980:
   1 million + 9 million + 2 million + 3 million = 15 million
   Total immigrants of Hispanic origin in 1990:
   1 million + 14 million + 3 million + 5 million = 23 million
   Difference: 23 million − 15 million = 8 million

9. The **population of immigrants of Hispanic origin in the United States** increased by a greater percent than the population of the United States between 1980 and 1990.
   U.S.A. population increase, 1980 to 1990: 9.7%
   Population in 1980: 227,000,000
   Population in 1990: 249,000,000
   Difference:
   249,000,000 − 227,000,000 = 22,000,000
   Percent of increase:
   22,000,000 ÷ 227,000,000 = .097 = 9.7%
   Hispanic immigrant population increase, 1980 to 1990: 53.3%
   Population in 1980: 15,000,000
   Population in 1990: 23,000,000
   Difference:
   23,000,000 − 15,000,000 = 8,000,000
   Percent of increase:
   8,000,000 ÷ 15,000,000 = .533 = 53.3%

10. There were about **2,300,000 marriages** in 1975.
    Marriages in 1970: 2,200,000
    Marriages in 1980: 2,400,000
    Difference: 2,400,000 − 2,200,000 = 200,000
    1975 is halfway between 1970 and 1980, so half the increase in the number of marriages probably occurred by 1975:
    200,000 × $\frac{1}{2}$ = 100,000
    Marriages in 1975:
    2,200,000 + 100,000 = 2,300,000
    (This question can also be answered by estimating from the graph. Find the point on the line that would represent 1975 and read across to the scale to find the number of marriages.)

11. The number of marriages in 1980 was 4% lower than the number of marriages in 1990.
    Number of marriages in 1980: 2,400,000
    Number of marriages in 1990: 2,500,000
    Difference: 2,500,000 − 2,400,000 = 100,000
    Percent decrease:
    100,000 ÷ 2,500,000 = .04 = 4%

12. **(2)** According to the trend the graph shows, the number of marriages in 2000 is likely to be between 2,500,000 and 2,599,999.
    The amount of change in the number of marriages has been decreasing since 1960.
    Between 1960 and 1970, the increase was 700,000.
    Between 1970 and 1980, the increase was 200,000.
    Between 1980 and 1990, the increase was 100,000.
    According to this trend of decreasing change, we can assume that between 1990 and 2000 the increase will be less than 100,000. Therefore, there will probably be only a few more marriages in 2000 than the 2,500,000 there were in 1990.

13. **72.6%** of children under 18 lived with both their parents in 1990.
    Sum of the other percents on the graph:
    21.6% + 3.1% + 2.2% + .5% = 27.4%
    Difference between that sum and the whole:
    100% − 27.4% = 72.6%

14. About **15,808,000** children lived with only one parent in 1990.
    Percent of children who lived with mother only: 21.6%
    Percent of children who lived with father only: 3.1%
    Total percent of children who lived with only one parent: 21.6% + 3.1% = 24.7%
    Number of children who lived with only one parent: 64,000,000 × .247 = 15,808,000

15. About **1,728,000** children lived with neither of their parents in 1990.
    Percent of children who lived with other relative(s): 2.2%
    Percent of children who lived with nonrelative(s): .5%
    Total percent of children who lived with neither parent: 2.2% + .5% = 2.7%
    Number of children who lived with neither parent: 64,000,000 × .027 = 1,728,000

GED PRACTICE 3 (page 208)

1. **(4)** The number 1,291,341,446 is in the *1983* row where it meets the *Quarters* column.

2. **(3)** Dimes minted in the United States from 1981 through 1990: 15,444,216,982
   Value: 15,444,216,982 × $.10
   = $1,544,421,698.20

3. **(1)** The number of pennies minted ranged from just under 9 billion in 1986 to nearly 17 billion in 1982. The number of nickels, dimes, and quarters minted ranged from less than 1 billion to slightly over 2 billion. Halves were minted only in the millions each year.

4. **(1)** The number of coins minted from 1981 through 1990 was 162,684,231,350. 10% of that number is 16,268,423,135, which is closest to the number of coins minted in 1981 (16,511,048,029).

5. **(2)** Total of all cars sold in the United States in 1990:
   U.S.A.   Japan   Germany
   7 million + $1\frac{1}{2}$ million + $\frac{1}{2}$ million
      Other   Total
   + $\frac{1}{2}$ million = $9\frac{1}{2}$ million

6. **(1)** 1985 U.S. sales of Japanese-made cars: 2,000,000
   1990 U.S. sales of Japanese-made cars: 1,500,000
   Difference:
   2,000,000 − 1,500,000 = 500,000
   Percent decrease:
   500,000 ÷ 2,000,000 = .25 = 25%

7. **(5)** Chicago's highest monthly normal temperature: 73° in July.
   Chicago's lowest monthly normal temperature: 21° in January.
   Difference: 73° − 21° = 52°

8. **(1)** From July through January, the trend of Chicago's monthly normal temperatures falls. After January, it begins to rise. From January through July, the trend of Chicago's monthly normal temperatures rises. After July, it begins to fall.

9. **(2)** 1991 U.S. livestock population: 180,000,000 (from the graph's footnote).
   At 33%, hogs and pigs made up about $\frac{1}{3}$ of the livestock population.
   So: 180,000,000 × $\frac{1}{3}$ = 60,000,000 hogs and pigs.

10. **(2)** 1991 U.S. livestock population: 180,000,000 (from the graph's footnote).
    Sheep and lambs made up 7% of the livestock population.
    So: 180,000,000 × .07 = 12,600,000 sheep and lambs
    And: 12,600,000 sheep and lambs × $65.60 per head = $826,560,000

POSTTEST (page 212)

1. **(2)** E. $\frac{1}{2} = \frac{16}{32}$; C. $\frac{5}{8} = \frac{20}{32}$; D. $\frac{3}{4} = \frac{24}{32}$; A. $\frac{25}{32} = \frac{25}{32}$; B. $\frac{13}{16} = \frac{26}{32}$

2. **(3)** $1\frac{1}{3}C + 2\frac{1}{3}C + 1\frac{2}{3}C = 4\frac{4}{3}C = 5\frac{1}{3}C$

3. **(4)** $\frac{1}{10}$ mi + $\frac{1}{4}$ mi + $\frac{2}{5}$ mi + $\frac{1}{2}$ mi + $\frac{3}{4}$ mi = $\frac{2}{20}$ mi + $\frac{5}{20}$ mi + $\frac{8}{20}$ mi + $\frac{10}{20}$ mi + $\frac{15}{20}$ mi = $\frac{40}{20}$ mi = 2 mi

4. **(2)** $12\frac{3}{4}$ gal − $7\frac{1}{3}$ gal = $12\frac{9}{12}$ gal − $7\frac{4}{12}$ gal = $5\frac{5}{12}$ gal

5. **(2)** $17\frac{1}{4}$ in. − $9\frac{7}{8}$ in. = $16\frac{10}{8}$ in. − $9\frac{7}{8}$ in. = $7\frac{3}{8}$ in.

6. **(4)** $325 × $\frac{3}{5}$ =
   $\dfrac{\overset{\$65}{\cancel{\$325}}}{1} \times \dfrac{3}{5} = \dfrac{\$195}{1} = \$195$

7. **(5)** $6\frac{1}{4}$ yd $\times 2\frac{2}{5}$ patterns per yd =
$\frac{\cancel{25}^5}{\cancel{4}_1} \times \frac{\cancel{12}^3}{\cancel{5}_1} = \frac{15}{1} = 15$ times

8. **(1)** Convert ounces to pounds: 1 lb = 16 oz
So: 2 oz ÷ 16 oz = $\frac{1}{8}$ lb

   Find the number of meatballs:
   $2\frac{1}{2}$ lb beef ÷ $\frac{1}{8}$ lb per meatball =
   $\frac{5}{2}$ lb ÷ $\frac{1}{8}$ lb =

   $\frac{5}{\cancel{2}_1} \times \frac{\cancel{8}^4}{1} = \frac{20}{1} = 20$ meatballs

   Divide by the number of people:
   20 meatballs ÷ 10 people
   = 2 meatballs each

9. **(4)** Perimeter = $8\frac{7}{10}$ in. × 3 sides
Perimeter = $\frac{87}{10} \times \frac{3}{1} = \frac{261}{10} = 26\frac{1}{10}$ in.

10. **(2)** Area = $\frac{\text{base} \times \text{height}}{2}$
Area = $\frac{8\frac{7}{10}\text{ in.} \times 7\frac{1}{2}\text{ in.}}{2}$
Area = $\frac{65\frac{1}{4}\text{ sq in.}}{2} = 32\frac{5}{8}$ sq in.

11. **(1)** Find the circumference of the garden:
Circumference = $\frac{22}{7} \times 17\frac{1}{2}$ ft

   Circumference = $\frac{22}{\cancel{7}_1} \times \frac{\cancel{35}^5}{\cancel{2}_1}^{11} = \frac{55}{1} = 55$ ft

   Divide the circumference by the amount of space between rose bushes:
   55 ft ÷ $2\frac{1}{2}$ ft per bush
   = $\frac{55}{1} \div \frac{5}{2}$
   = $\frac{\cancel{55}^{11}}{1} \times \frac{2}{\cancel{5}_1} = \frac{22}{1} = 22$ rose bushes

12. **(1)** Find the area of the square:
Area = length × width
Area = 2 ft × 2 ft = 4 sq ft

   Find the area of the circle:
   (The diameter is 2 ft, so the radius is 1 ft.)
   Area = $\frac{22}{7}$ × radius × radius
   Area = $\frac{22}{7}$ × 1 ft × 1 ft
   Area = $\frac{22}{7} \times \frac{1}{1} \times \frac{1}{1} = \frac{22}{7} = 3\frac{1}{7}$ sq ft

   Subtract the area of the circle from the area of the square:
   4 sq ft − $3\frac{1}{7}$ sq ft = $\frac{6}{7}$ sq ft

13. **(5)** Change $12.75 to a mixed number:
$12.75 = 12\frac{3}{4}$

   Multiply:
   $10\frac{2}{3}$ boxes of tiles × 12\frac{3}{4}$ per box
   = $\frac{\cancel{32}^8}{\cancel{3}_1} \times \frac{\cancel{51}^{17}}{\cancel{4}_1} = \frac{136}{1}$ = $136 (or $136.00)

14. **(5)** The value of 2% is not equal to the values of the other choices:
2% = .02 = $\frac{2}{100} = \frac{1}{50}$

15. **(4)** The percent equivalent of $\frac{3}{8}$ is 37.5% or $37\frac{1}{2}$%.

16. **(2)** part = % × whole
part = $12\frac{1}{2}$% × $96,000
part = .125 × $96,000 (or, $\frac{1}{8}$ × $96,000)
part = $12,000

17. **(4)** % = part ÷ whole
% = $\frac{\$245}{\$350} = \frac{7}{10}$ (or, $245 ÷ $350 = .7)
% = 70%

18. **(5)** whole = part ÷ percent
whole = $157.50 ÷ 4.5%
whole = $157.50 ÷ .045
whole = $3500 (or $3500.00)

19. **(2)** Find the percent that represents the TV's sale price:
100% − 20% = 80%

   To estimate the sale price of the TV, first round the original price:
   $79.98 can be rounded to $80.

   Solve the problem:
   part = % × whole
   part = 80% × $80
   part = .8 × $80 (or $\frac{4}{5}$ × $80)
   part = $64

20. **(1)** Find the part of the trip Janice drives the second day:
400 mi − 350 mi = 50 mi

   Find what percent of her trip 50 mi is:
   % = part ÷ whole
   % = $\frac{50}{400} = \frac{1}{8}$
   % = $12\frac{1}{2}$%

21. **(4)** Your account balance would include the amount deposited (100% × $1200) plus the interest ($5\frac{1}{2}$% × $1200).

    Add the two percents:
    $100\% + 5\frac{1}{2}\% = 105\frac{1}{2}\%$

    Find the account balance:
    part = % × whole
    part = $105\frac{1}{2}$% × $1200
    part = 1.055 × $1200
    part = $1266

22. **(3)** Find the percent that represents the size of the class after six weeks:
    100% − 20% = 80%

    Find the number of students still taking the class:
    part = % × whole
    part = 80% × 30 students
    part = .8 × 30
    part = 24 students

23. **(2)** Find the amount of the markup:
    $1.56 − $1.50 = $.06

    Find the percent of the markup:
    % = part ÷ whole
    % = $.06 ÷ $1.50 = .04
    % = 4%

24. **(5)** U.S. manufacturing workers worked about 1950 hours in 1990.
    German manufacturing workers worked about 1550 hours in 1990.
    The difference:
        1950 hours − 1550 hours = 400 hours

25. **(2)** The point on the *Current dollars* line that stands for 1984 is even with $15,000 on the vertical axis.

26. **(1)** The points that stand for 1970 and 1990 on the *1990 dollars* line fall between labels on the vertical axis, so you need to estimate:

    In 1990 dollars, the median income of workers was $21,000 in 1970.
    In 1990 dollars, the median income of workers was $18,000 in 1990.

    Find the difference:
    $21,000 (in 1970) − $18,000 (in 1990)
    = $3,000 less in 1990

27. **(1)** Group 1 received 44.3%—nearly 50%, or half—of all personal income in 1990. Notice that the *Group 1* section of the graph covers almost half of the circle.

28. **(3)** Find the amount of Group 1 families' 1990 income:
    $4645.5 billion × 44.3% =
    $4645.5 billion × .443 = $2057.9565 billion

    Find the amount of Group 5 families' 1990 income:
    $4645.5 billion × 4.5% =
    $4645.5 billion × .045 = $209.0385 billon

    Find the difference:
    $2057.9565 billion − $209.0385 billion
    = $1848.918 billion